Answer

SRA Corrective Mathem

Division

A Direct Instruction Program

Siegfried Engelmann • Doug Carnine

SRA
Columbus, OH

The McGraw·Hill Companies

Cover and title page photo credits:
©Photodisc/Getty Images, Inc.

SRAonline.com

McGraw Hill SRA

Send all inquiries to:
SRA/McGraw-Hill
4400 Easton Commons
Columbus, OH 43219

Printed in the United States of America.

ISBN 0-07-602471-7

6 7 8 9 10 11 GLO 15 14 13 12 11 10 09

The McGraw·Hill Companies

Answer Key
Division Contents

Division Preskill Test

A Subtraction

60 − 32 = 28	486 − 189 = 297	307 − 245 = 62	402 − 386 = 16	452 − 180 = 272
356 − 186 = 170	500 − 364 = 136	454 − 327 = 127	478 − 296 = 182	375 − 278 = 97

B Multiplication

35 × 8 = 280	54 × 7 = 378	46 × 6 = 276	54 × 4 = 216	34 × 3 = 102
94 × 9 = 846	75 × 6 = 450	95 × 8 = 760	37 × 7 = 259	85 × 6 = 510

Division Preskill Test

A Subtraction

60 − 32 = 28	486 − 189 = 297	307 − 245 = 62	402 − 386 = 16	452 − 180 = 272
356 − 186 = 170	500 − 364 = 136	454 − 327 = 127	478 − 296 = 182	375 − 278 = 97

B Multiplication

35 × 8 = 280	54 × 7 = 378	46 × 6 = 276	54 × 4 = 216	34 × 3 = 102
94 × 9 = 846	75 × 6 = 450	95 × 8 = 760	37 × 7 = 259	85 × 6 = 510

same as page a

Division Placement Test

A

5 ⟌ 15 = 3	1 ⟌ 7 = 7	3 ⟌ 6 = 2	9 ⟌ 36 = 4	1 ⟌ 8 = 8
9 ⟌ 18 = 2	3 ⟌ 12 = 4	1 ⟌ 3 = 3	5 ⟌ 25 = 5	5 ⟌ 5 = 1
3 ⟌ 6 = 2	9 ⟌ 9 = 1	5 ⟌ 10 = 2	9 ⟌ 27 = 3	3 ⟌ 9 = 3
5 ⟌ 20 = 4	9 ⟌ 45 = 5	1 ⟌ 4 = 4	3 ⟌ 3 = 1	3 ⟌ 15 = 5

B

5 ⟌ 70 = 14	9 ⟌ 216 = 24	5 ⟌ 116 = 23	3 ⟌ 104 = 34	3 ⟌ 45 = 15
−5	−18	−10	−9	−3
20	36	16	14	15
−20	−36	−15	−12	−15
0	0	1	2	0

3 buses left Midville each day. 12 buses left in all. How many days did buses leave Midville?

3 ⟌ 12 = 4

4 days

Every time Fred went jogging, he ran 5 blocks. He ran 10 blocks. How many times did he go jogging?

5 ⟌ 10 = 2

2 times

Part B continues on the next page.

Division Placement Test (continued)

Pete did 10 problems on each page. He did 5 pages. How many problems did he do?

[50] problems

$$\begin{array}{r} 10 \\ \times\ 5 \\ \hline 50 \end{array}$$

Betty typed 2 pages each hour. She typed 8 pages. How many hours did she type?

[4] hours

2)8 = 4

There are 6 apples in each pile. There are 2 piles of apples. How many apples in all?

[12] apples

$$\begin{array}{r} 6 \\ \times\ 2 \\ \hline 12 \end{array}$$

C

48)264 = 5 −240 24	27)162 = 6 −162 0	82)2354 = 28 −164 714 −656 58
54)3267 = 60 −324 27	74)7934 = 107 −74 534 −518 16	73)7142 = 97 −657 572 −511 61

Lesson 1

Bonus [ ]

1

A 2)6 = 3 B 5)10 = 2 C 2)12 = 6

2

A 5)15 [ ] B 5)10 [ ] C 5)20 [ ] D 5)5 [ ]

3

A 9)45 [ ] B 9)9 [ ] C 9)27 [ ] D 9)18 [ ]

4

A 4)24 = 6 4)24 = 6 6)24 = 4 4)24 = 6 6)24 = 4

B 3)15 = 5 3)15 = 5 5)15 = 3 C 4)12 = 3 4)12 = 3 3)12 = 4

5

A 5)25 = 5 B 9)18 = 2 C 9)27 = 3 D 5)15 = 3

Lesson 2

Bonus [ ]

1

A 5)25 [ ] B 5)10 [ ] C 5)20 [ ] D 5)15 [ ]

2

A 9)36 [ ] B 9)18 [ ] C 9)45 [ ] D 9)27 [ ]

3

A 5)20 = 4 B 5)10 = 2 C 5)25 = 5

4

A 5)20 = 4 5)20 = 4 4)20 = 5 B 9)36 = 4 9)36 = 4 4)36 = 9

C 9)18 = 2 9)18 = 2 2)18 = 9 D 5)10 = 2 5)10 = 2 2)10 = 5

Lesson 3

Bonus [ ]

1

A 9)45 [ ] B 9)9 [ ] C 9)27 [ ] D 9)18 [ ]

2

A 5)25 [ ] B 5)10 [ ] C 5)5 [ ] D 5)20 [ ]

3

A 9)27 = 3 B 5)15 = 3 C 9)18 = 2

4

A 9)18 = 2 9)18 = 2 2)18 = 9 B 9)36 = 4 9)36 = 4 4)36 = 9

C 5)15 = 3 5)15 = 3 3)15 = 5 D 5)5 = 1 5)5 = 1 1)5 = 5

5

A 5)5 = [1] 5)5 = 1 1)5 = 5 B 5)10 = [2] 5)10 = 2 2)10 = 5

C 5)15 = [3] 5)15 = 3 3)15 = 5 D 5)20 = [4] 5)20 = 4 4)20 = 5

E 5)25 = [5] 5)25 = 5 5)25 = 5

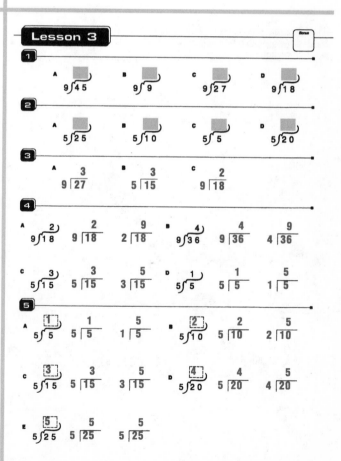

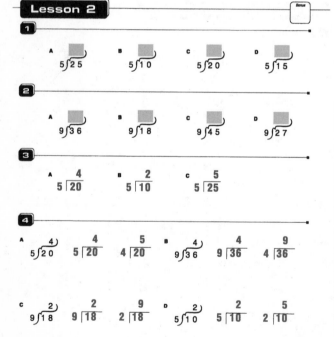

Lesson 4

Bonus

1

A 5)5 = [1] 5)5 = 1 1)5 = 5
B 5)10 = [2] 5)10 = 2 2)10 = 5
C 5)15 = [3] 5)15 = 3 3)15 = 5
D 5)20 = [4] 5)20 = 4 4)20 = 5

2

A 1)4 = [4] 1)4 = 4 4)4 = 1
B 1)10 = [10] 1)10 = 10 10)10 = 1
C 1)7 = [7] 1)7 = 7 7)7 = 1
D 1)5 = [5] 1)5 = 5 5)5 = 1

3

A 2)6 = 3 B 5)10 = 2 C 2)12 = 6 D 6)18 = 3 E 7)14 = 2

4

A 1)7 B 1)4 C 1)2 D 1)6

Lesson 5

Bonus

1

A 5)25 = [5] 5)25 = 5 5)25 = 5
B 5)10 = [2] 5)10 = 2 2)10 = 5
C 5)5 = [1] 5)5 = 1 1)5 = 5
D 5)20 = [4] 5)20 = 4 4)20 = 5

2

A 5)10 = 2 B 5)20 = 4 C 5)5 = 1 D 5)15 = 3

3

A 1)7 = [7] 1)7 = 7 7)7 = 1
B 1)4 = [4] 1)4 = 4 4)4 = 1
C 1)9 = [9] 1)9 = 9 9)9 = 1
D 1)6 = [6] 1)6 = 6 6)6 = 1

4

A 8)16 = 2 B 2)10 = 5 C 4)12 = 3 D 9)9 = 1 E 2)8 = 4

5

A 1)10 B 1)8 C 1)1 D 1)9

Lesson 6

Facts + Bonus = TOTAL

1

A 1)3 = [3] 1)3 = 3 3)3 = 1
B 1)10 = [10] 1)10 = 10 10)10 = 1
C 1)8 = [8] 1)8 = 8 8)8 = 1
D 1)1 = [1] 1)1 = 1 1)1 = 1

2

A 1)3 = 3 B 1)6 = 6 C 1)2 = 2 D 1)9 = 9

3

(0) 1 2 3 4 (5) 6 7 8 9 (10) 11 12 13 14 (15) 16 17 18 19 (20) 21 22 23 24 (25)

4

5)15 = 3 5)16 = 3 with R 5)17 = 3 with R 5)18 = 3 with R 5)19 = 3 with R

5

A 9)27 B 9)36 C 9)18 D 9)45

6

A 5)25 = [5] 5)25 = 5 5)25 = 5
B 5)15 = [3] 5)15 = 3 3)15 = 5
C 5)5 = [1] 5)5 = 1 1)5 = 5
D 5)10 = [2] 5)10 = 2 2)10 = 5

Lesson 7

Facts + Bonus = TOTAL

1

A 9)9 = [1] 9)9 = 1 1)9 = 9
B 9)18 = [2] 9)18 = 2 2)18 = 9
C 9)27 = [3] 9)27 = 3 3)27 = 9
D 9)36 = [4] 9)36 = 4 4)36 = 9

2

(0) 1 2 3 4 (5) 6 7 8 9 (10) 11 12 13 14 (15) 16 17 18 19 (20) 21 22 23 24 (25)

3

5)10 = 2 5)11 = 2 with R 5)12 = 2 with R 5)13 = 2 with R 5)14 = 2 with R

4

A 9)18 B 9)9 C 9)27 D 9)36

5

A 1)7 = [7] 1)7 = 7 7)7 = 1
B 1)10 = [10] 1)10 = 10 10)10 = 1
C 1)4 = [4] 1)4 = 4 4)4 = 1
D 1)5 = [5] 1)5 = 5 5)5 = 1

Lesson 8

Facts + Problems + Bonus = TOTAL

1

A. 9)9 → [1]; 9)9 → 1; 1)9 → 9
B. 9)18 → [2]; 9)18 → 2; 2)18 → 9
C. 9)27 → [3]; 9)27 → 3; 3)27 → 9
D. 9)36 → [4]; 9)36 → 4; 4)36 → 9

2

A. 5)21 → ▢; 5)20 → 4
B. 5)12 → ▢; 5)10 → 2
C. 5)16 → ▢; 5)15 → 3
D. 5)9 → ▢; 5)5 → 1
E. 5)19 → ▢; 5)15 → 3

3

A. 9)36 → ▢
B. 9)45 → ▢
C. 9)18 → ▢
D. 9)27 → ▢

4

A. 5)5 → [1]; 5)5 → 1; 1)5 → 5
B. 5)20 → [4]; 5)20 → 4; 4)20 → 5
C. 5)15 → [3]; 5)15 → 3; 3)15 → 5
D. 5)10 → [2]; 5)10 → 2; 2)10 → 5

5

Write all the numbers 5 goes into 3 times **with** a remainder.

5)15 → 3
A. 5)16 → 3 R
B. 5)17 → 3 R
C. 5)18 → 3 R
D. 5)19 → 3 R

Lesson 9

Problems + Bonus = TOTAL

1

A. 9)27 → [3]; 9)27 → 3; 3)27 → 9
B. 9)9 → [1]; 9)9 → 1; 1)9 → 9
C. 9)18 → [2]; 9)18 → 2; 2)18 → 9
D. 9)45 → [5]; 9)45 → 5; 5)45 → 9

2

A. 5)11 → ▢; 5)10 → 2
B. 5)22 → ▢; 5)20 → 4
C. 5)7 → ▢; 5)5 → 1
D. 5)13 → ▢; 5)10 → 2
E. 5)17 → ▢; 5)15 → 3

3

A. 5)21 → 4; −20; 1
B. 5)23 → 4; −20; 3
C. 5)17 → 3; −15; 2
D. 5)14 → 2; −10; 4
E. 5)18 → 3; −15; 3
F. 5)12 → 2; −10; 2

Lesson 9 (continued)

4

A. 5)22 → 4; −20; 2
B. 5)11 → 2; −10; 1
C. 5)23 → 4; −20; 3
D. 5)19 → 3; −15; 4
E. 5)8 → 1; −5; 3
F. 5)16 → 3; −15; 1
G. 5)12 → 2; −10; 2
H. 5)24 → 4; −20; 4

5

Write all the numbers 5 goes into 4 times **with** a remainder.

5)20 → 4
A. 5)21 → 4 R
B. 5)22 → 4 R
C. 5)23 → 4 R
D. 5)24 → 4 R

Write all the numbers 5 goes into 2 times **with** a remainder.

5)10 → 2
E. 5)11 → 2 R
F. 5)12 → 2 R
G. 5)13 → 2 R
H. 5)14 → 2 R

Lesson 10

Facts + Problems + Bonus = TOTAL

1

A. 9)18 → [2]; 9)18 → 2; 2)18 → 9
B. 9)45 → [5]; 9)45 → 5; 5)45 → 9
C. 9)36 → [4]; 9)36 → 4; 4)36 → 9
D. 9)9 → [1]; 9)9 → 1; 1)9 → 9

2

A. 9)27 → 3
B. 9)45 → 5
C. 9)18 → 2
D. 9)36 → 4

3

A. 5)16 → ▢; 5)15 → 3
B. 5)24 → ▢; 5)20 → 4
C. 5)13 → ▢; 5)10 → 2
D. 5)18 → ▢; 5)15 → 3
E. 5)8 → ▢; 5)5 → 1

4

A. 5)17 → 3; −15; 2
B. 5)14 → 2; −10; 4
C. 5)27 → 5; −25; 2
D. 5)8 → 1; −5; 3
E. 5)22 → 4; −20; 2

Lesson 10 (continued)

5

	Problem	Answer	Subtract	Remainder
A	5)9	1	−5	4
B	5)13	2	−10	3
C	5)24	4	−20	4
D	5)6	1	−5	1
E	5)27	5	−25	2
F	5)23	4	−20	3
G	5)19	3	−15	4
H	5)12	2	−10	2

6

5)15 = 3   9)18 = 2   9)45 = 5   1)10 = 10   5)25 = 5   9)36 = 4   9)9 = 1

9)36 = 4   1)7 = 7   5)20 = 4   5)15 = 3   1)2 = 2   5)5 = 1   5)25 = 5

7

Write all the numbers 5 goes into 1 time with a remainder.

5)5 = 1   A 5)6 = 1 R   B 5)7 = 1 R   C 5)8 = 1 R   D 5)9 = 1 R

Write all the numbers 5 goes into 4 times with a remainder.

5)20 = 4   E 5)21 = 4 R   F 5)22 = 4 R   G 5)23 = 4 R   H 5)24 = 4 R

Lesson 11

Facts + Problems + Bonus = TOTAL

1

(0) 1 2 3 4 5 6 7 8 (9) 10 11 12 13 14 15 16 17 (18) 19 20 21 22 23 24 25 26 (27)

2

9)18 = 2   9)19 = 2 R   9)20 = 2 R   9)21 = 2 R   9)22 = 2 R

9)23 = 2 R   9)24 = 2 R   9)25 = 2 R   9)26 = 2 R

3

A 5)17 ▢   5)15 = 3   B 5)8 ▢   5)5 = 1   C 5)23 ▢   5)20 = 4

D 5)9 ▢   5)5 = 1   E 5)12 ▢   5)10 = 2   F 5)19 ▢   5)15 = 3

4

	Problem	Answer	Subtract	Remainder
A	5)18	3	−15	3
B	9)19	2	−18	1
C	5)17	3	−15	2
D	9)6	0	−0	6
E	9)38	4	−36	2
F	5)6	1	−5	1

5

	Problem	Answer	Subtract	Remainder
A	9)22	2	−18	4
B	9)33	3	−27	6
C	9)44	4	−36	8

Lesson 11 (continued)

6

A The big number in a times problem is 10. One small number is 5.

B One small number in a times problem is 10. The other small number is 5.

C One small number in a times problem is 9. The other small number is 4.

D The big number in a times problem is 36. One small number is 9.

E The big number in a times problem is 20. One small number is 5.

7

5)10 = 2   9)18 = 2   9)9 = 1   1)8 = 8   9)36 = 4   5)25 = 5   5)15 = 3

9)27 = 3   5)5 = 1   1)4 = 4   5)10 = 2   1)6 = 6   5)20 = 4   9)18 = 2

8

Finish working these problems and show the remainders.

	Problem	Answer	Subtract	Remainder
A	9)26	2	−18	8
B	5)16	3	−15	1
C	5)29	5	−25	4
D	9)12	1	−9	3
E	5)12	2	−10	2
F	5)8	1	−5	3
G	9)38	4	−36	2
H	5)29	5	−25	4
I	5)18	3	−15	3
J	9)47	5	−45	2

Lesson 12

Test + Problems + Bonus = TOTAL

1

(0) 1 2 3 4 5 6 7 8 (9) 10 11 12 13 14 15 16 17 (18) 19 20 21 22 23 24 25 26 (27) 28 29 30 31 32 33 34 35 (36) 37 38 39 40 41 42 43 44 (45) 46 47 . . .

2

9)27 = 3   9)28 = 3 R   9)29 = 3 R   9)30 = 3 R   9)31 = 3 R

9)32 = 3 R   9)33 = 3 R   9)34 = 3 R   9)35 = 3 R

3

A 5)21 ▢   5)20 = 4   B 5)24 ▢   5)20 = 4   C 5)13 ▢   5)10 = 2

D 5)18 ▢   5)15 = 3   E 5)6 ▢   5)5 = 1   F 5)17 ▢   5)15 = 3

4

	Problem	Answer	Subtract	Remainder
A	9)38	4	−36	2
B	5)14	2	−10	4
C	5)23	4	−20	3
D	9)17	1	−9	8
E	9)29	3	−27	2

5

	Problem	Answer	Subtract	Remainder
A	9)52	5	−45	7
B	9)23	2	−18	5
C	9)40	4	−36	4

Lesson 12 (continued)

6

A One small number in a times problem is 5. The other small number is 3.

B The big number in a times problem is 45. One small number is 9.

7

A The big number in a times problem is 20. One small number is 5. What's the third number? — 5)20 = 4

B One small number in a times problem is 3. The other small number is 4. What's the third number? — 3 × 4 = 12

C One small number in a times problem is 1. The other small number is 8. What's the third number? — 1 × 8 = 8

D The big number in a times problem is 18. One small number is 9. What's the third number? — 9)18 = 2

8

5)5 = 1	9)27 = 3	9)36 = 4	5)15 = 3	9)9 = 1	9)45 = 5	1)6 = 6
1)1 = 1	9)18 = 2	5)10 = 2	1)9 = 9	5)25 = 5	9)36 = 4	5)20 = 4
9)27 = 3	5)15 = 3	9)18 = 2	5)20 = 4	9)45 = 5	5)25 = 5	1)8 = 8

9

Finish working these problems and show the remainders.

A	B	C	D	E
5)8 = 1	9)38 = 4	5)29 = 5	5)18 = 3	9)47 = 5
−5	−36	−25	−15	−45
3	2	4	3	2

Lesson 13

Facts + Problems + Bonus = TOTAL

1

A 9)19 [shaded]	9)18 = 2
B 9)30 [shaded]	9)27 = 3
C 9)12 [shaded]	9)9 = 1
D 9)24 [shaded]	9)18 = 2
E 9)41 [shaded]	9)36 = 4

2

A	B	C	D	E	F
~~5)18 = 4, −20~~	~~9)41 = 5, −45~~	~~5)10 = 3, −15~~	9)29 = 3	~~9)30 = 4, −36~~	5)14 = 2
			−27		−10
			2		4

3

A One small number in a times problem is 6. The other small number is 3. What's the third number? — 6 × 3 = 18

B One small number in a times problem is 2. The other small number is 8. What's the third number? — 2 × 8 = 16

C The big number in a times problem is 27. One small number is 9. What's the third number? — 9)27 = 3

D The big number in a times problem is 15. One small number is 5. What's the third number? — 5)15 = 3

E The big number in a times problem is 36. One small number is 9. What's the third number? — 9)36 = 4

F One small number in a times problem is 7. The other small number is 5. What's the third number? — 7 × 5 = 35

Lesson 13 (continued)

4

5)20 = 4	9)27 = 3	1)4 = 4	5)5 = 1	9)18 = 2	5)10 = 2	5)15 = 3
9)9 = 1	9)36 = 4	5)10 = 2	9)45 = 5	5)15 = 3	9)18 = 2	9)36 = 4

5

Finish working these problems and show the remainders.

A	B	C
4)23 = 5	9)49 = 5	3)28 = 9
−20	−45	−27
3	4	1

6

Write all the numbers 9 goes into 3 times with a remainder.

9)27 = 3

A	B	C	D
9)28 = 3 R	9)29 = 3 R	9)30 = 3 R	9)31 = 3 R
E 9)32 = 3 R	F 9)33 = 3 R	G 9)34 = 3 R	H 9)35 = 3 R

7

Write the facts with no remainders.

A 5)6 [shaded]	5)5 = 1
B 5)17 [shaded]	5)15 = 3
C 5)12 [shaded]	5)10 = 2
D 5)23 [shaded]	5)20 = 4
E 5)14 [shaded]	5)10 = 2
F 5)19 [shaded]	5)15 = 3

Lesson 13 (continued)

8

Write the answer. Then multiply and subtract to find the remainder.

A	B	C	D	E
5)17 = 3	9)31 = 3	9)20 = 2	5)8 = 1	9)14 = 1
−15	−27	−18	−5	−9
2	4	2	3	5

F	G	H	I	J
9)21 = 2	5)9 = 1	5)16 = 3	9)40 = 4	9)19 = 2
−18	−5	−15	−36	−18
3	4	1	4	1

9

A	5)5 = [1]	5)5 = 1	1)5 = 5
B	5)20 = [4]	5)20 = 4	4)20 = 5
C	5)15 = [3]	5)15 = 3	3)15 = 5
D	5)10 = [2]	5)10 = 2	2)10 = 5
E	9)18 = [2]	9)18 = 2	2)18 = 9
F	9)45 = [5]	9)45 = 5	5)45 = 9
G	9)36 = [4]	9)36 = 4	4)36 = 9
H	9)9 = [1]	9)9 = 1	1)9 = 9

Lesson 14

Facts + Problems + Bonus = TOTAL

1

A 3)15 [ ] B 3)9 [ ] C 3)6 [ ] D 3)12 [ ]

2

A 3)3 = [1]; 3)3 = 1; 1)3 = 3
B 3)6 = [2]; 3)6 = 2; 2)6 = 3
C 3)9 = [3]; 3)9 = 3
D 3)12 = [4]; 3)12 = 4; 4)12 = 3

3

A 9)21 [ ]; 9)18 = 2
B 9)12 [ ]; 9)9 = 1
C 9)39 [ ]; 9)36 = 4
D 9)32 [ ]; 9)27 = 3
E 9)11 [ ]; 9)9 = 1

4

A 9)19 = 3, −27 (crossed out)
B 9)43 = 5, −45 (crossed out)
C 5)23 = 4, −20, remainder 3
D 9)16 = 2, −18 (crossed out)
E 5)26 = 5, −25, remainder 1
F 5)11 = 3, −15 (crossed out)

Lesson 14 (continued)

5

A There are 4 blocks in each pile. There are 16 blocks in all.

B Every time Sam goes to the store, he buys 6 carrots. He has 42 carrots in all.

C There are 9 towels in each box. There are 27 towels in all.

D Mary Jo builds 4 benches each day that she works. She builds 20 benches in all.

E Every time Mrs. Whitehead goes jogging, she runs 5 kilometers. She runs 35 kilometers in all.

6

3)12 = 4; 5)25 = 5; 3)15 = 5; 9)45 = 5; 3)9 = 3; 9)18 = 2; 3)12 = 4

3)9 = 3; 3)15 = 5; 9)36 = 4; 5)20 = 4; 3)12 = 4; 9)27 = 3; 3)9 = 3

7

Write all the numbers 9 goes into 4 times with a remainder.

9)36 = 4

A 9)37 = 4 R B 9)38 = 4 R C 9)39 = 4 R D 9)40 = 4 R

E 9)41 = 4 R F 9)42 = 4 R G 9)43 = 4 R H 9)44 = 4 R

Lesson 14 (continued)

8

Write the facts with no remainders.

A 5)14 [ ]; 5)10 = 2
B 5)16 [ ]; 5)15 = 3
C 5)19 [ ]; 5)15 = 3
D 5)24 [ ]; 5)20 = 4
E 5)11 [ ]; 5)10 = 2
F 5)8 [ ]; 5)5 = 1

9

Write the fact for each problem.

A One small number in a times problem is 2. The other small number is 4. What's the third number? — $2 \times 4 = 8$

B The big number in a times problem is 18. One small number is 9. What's the third number? — 9)18 = 2

C The big number in a times problem is 20. One small number is 5. What's the third number? — 5)20 = 4

D One small number in a times problem is 2. The other small number is 5. What's the third number? — $2 \times 5 = 10$

10

Write the answer. Then multiply and subtract to find the remainder.

A 9)40 = 4, −36, remainder 4
B 5)16 = 3, −15, remainder 1
C 9)21 = 2, −18, remainder 3
D 5)9 = 1, −5, remainder 4
E 9)19 = 2, −18, remainder 1

Lesson 15

Facts + Problems + Bonus = TOTAL

1

A 3)6 [ ] B 3)15 [ ] C 3)3 [ ] D 3)12 [ ]

2

A 3)15 = [5]; 3)15 = 5; 5)15 = 3
B 3)6 = [2]; 3)6 = 2; 2)6 = 3
C 3)9 = [3]; 3)9 = 3
D 3)12 = [4]; 3)12 = 4; 4)12 = 3

3

A 9)29 [ ]; 9)27 = 3
B 9)31 [ ]; 9)27 = 3
C 9)13 [ ]; 9)9 = 1
D 9)38 [ ]; 9)36 = 4
E 9)20 [ ]; 9)18 = 2

4

A 9)21 = ~~3~~ 2, −36; 9)21 = 2, −18, remainder 3
B 5)23 = ~~5~~ 4, −25; 5)23 = 4, −20, remainder 3
C 9)22 = ~~3~~ 2, −27; 9)22 = 2, −18, remainder 4
D 5)14 = ~~4~~ 2, −20; 5)14 = 2, −10, remainder 4
E 9)40 = ~~5~~ 4, −45; 9)40 = 4, −36, remainder 4
F 9)13 = ~~2~~ 1, −18; 9)13 = 1, −9, remainder 4

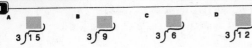

Lesson 15 (continued)

5

A There are 5 paper cups in each pile. There are 25 paper cups in all.

B Every time Ned goes to the store, he buys 3 oranges. He has 12 oranges in all.

6

A Every time Carlos walked to school, he walked 5 blocks. Carlos walked 200 blocks in all. How many times did Carlos walk to school?

B Every time Carlos walked to school, he walked 5 blocks. Carlos walked to school 200 times. How many blocks did Carlos walk in all?

C Every time Pam's team scored, they got 7 points. Pam's team scored 75 times. How many points did they get in all?

D Every time Yuki used pens, she used 4 pens. Yuki used 12 pens in all. How many times did Yuki use pens?

7

4	3	3	5	7	4	5
3)12	5)15	3)9	3)15	1)7	3)12	5)25
1	5	4	10	1	5	3
5)5	3)15	3)12	1)10	9)9	3)15	3)9

8

Write the fact for each problem.

A The big number in a times problem is 20. One small number is 5. What's the third number?
4
5)20

B One small number in a times problem is 5. The other small number is 4. What's the third number?
5
× 4
20

C The big number in a times problem is 45. The other number is 9. What's the third number?
5
9)45

Lesson 15 (continued)

9

Write all the numbers 9 goes into 5 times with a remainder.

	A	B	C	D
5	5 R	5 R	5 R	5 R
9)45	9)46	9)47	9)48	9)49

E	F	G	H
5 R	5 R	5 R	5 R
9)50	9)51	9)52	9)53

10

Write the facts with no remainders.

A [ ]	3	**B** [ ]	2	**C** [ ]	1
5)18	5)15	5)11	5)10	5)7	5)5
D [ ]	4	**E** [ ]	2	**F** [ ]	3
5)21	5)20	5)12	5)10	5)17	5)15

11

Write the answer. Then multiply and subtract to find the remainder.

A	B	C	D	E
3	2	1	4	2
5)18	9)20	5)6	9)41	9)22
−15	−18	−5	−36	−18
3	2	1	5	4

Lesson 16

Facts + Problems + Bonus = TOTAL

1

A [2]	2	3	**B** [5]	5	3
3∫6	3)6	2)6	3∫15	3)15	5)15
C [1]	1	3	**D** [3]	3	
3∫3	3)3	1)3	3∫9	3)9	

2

A [ ]	2	**B** [ ]	3	**C** [ ]	2
9)20	9)18	9)29	9)27	9)22	9)18
D [ ]	1	**E** [ ]	4	**F** [ ]	1
9)14	9)9	9)39	9)36	9)15	9)9

3

A 2 / ~~3~~	2	**B** 1 / ~~2~~	1	**C** 2 / ~~3~~	2
5)13	5)13	9)16	9)16	5)11	5)11
−15	−10	−27	− 9	−15	−10
	3		7		1
D 3 / ~~4~~	3	**E** 2 / ~~3~~	2	**F** 1 / ~~2~~	1
5)18	5)18	9)23	9)23	9)17	9)17
−20	−15	−27	−18	−18	− 9
	3		5		8

Lesson 16 (continued)

4

A Every day Mattie read 3 books. Mattie read 18 books in all. How many days did Mattie read books?
3)18

B Every time Jack fed the dogs, he used 7 cans of food. He fed the dogs 28 times. How many cans of food did Jack use in all?
28
× 7

C Every week Ron ate 5 bananas. Ron ate bananas for 50 weeks. How many bananas did Ron eat in all?
50
× 5

D Every time Ann made a dress, she used 3 meters of cloth. Ann used 45 meters of cloth in all. How many dresses did Ann make?
3)45

5

A Don made 40 sandwiches each week. He made sandwiches for 13 weeks.

B Each day Mary planted 6 trees. Mary has planted trees for 13 days.

C Kit loaded 6 trucks every day. She loaded trucks for 13 days.

D Don washed 10 tables each hour. He washed tables for 5 hours.

E Every afternoon that Peg looked for worms, she found 12. Peg looked for worms for 5 afternoons.

6

2	2	4	2	3	2	5
9)18	3)6	3)12	5)10	3)9	3)6	5)25

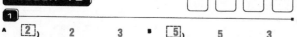

Lesson 16 (continued)

7 Write the fact for each problem.

A The big number in a times problem is 20. One small number is 5. What's the third number?

$$\dfrac{4}{5\,\overline{)20}}$$

B One small number in a times problem is 5. The other small number is 5. What's the third number?

$$\begin{array}{r} 5 \\ \times\,5 \\ \hline 25 \end{array}$$

C The big number in a times problem is 30. One small number is 5. What's the third number?

$$\dfrac{5}{6\,\overline{)30}}$$

D The big number in a times problem is 18. One small number is 9. What's the third number?

$$\dfrac{2}{9\,\overline{)18}}$$

8

A $\begin{array}{r}1\\5\overline{)7}\\-5\\\hline 2\end{array}$ B $\begin{array}{r}4\\9\overline{)40}\\-36\\\hline 4\end{array}$ C $\begin{array}{r}2\\9\overline{)19}\\-18\\\hline 1\end{array}$ D $\begin{array}{r}5\\5\overline{)28}\\-25\\\hline 3\end{array}$ E $\begin{array}{r}1\\9\overline{)15}\\-9\\\hline 6\end{array}$

F $\begin{array}{r}3\\5\overline{)17}\\-15\\\hline 2\end{array}$ G $\begin{array}{r}3\\9\overline{)30}\\-27\\\hline 3\end{array}$ H $\begin{array}{r}1\\5\overline{)6}\\-5\\\hline 1\end{array}$ I $\begin{array}{r}5\\9\overline{)46}\\-45\\\hline 1\end{array}$ J $\begin{array}{r}2\\9\overline{)20}\\-18\\\hline 2\end{array}$

K $\begin{array}{r}1\\9\overline{)14}\\-9\\\hline 5\end{array}$ L $\begin{array}{r}4\\5\overline{)24}\\-20\\\hline 4\end{array}$ M $\begin{array}{r}4\\9\overline{)38}\\-36\\\hline 2\end{array}$ N $\begin{array}{r}5\\5\overline{)27}\\-25\\\hline 2\end{array}$ O $\begin{array}{r}2\\5\overline{)11}\\-10\\\hline 1\end{array}$

Lesson 17

[Facts + Problems + Bonus = TOTAL]

1

A [▨] 5)35   B [▨] 5)50   C [▨] 5)30   D [▨] 5)45

2

A [6] 5)30, 6 5)30, 5 6)30   B [7] 5)35, 7 5)35, 5 7)35

C [8] 5)40, 8 5)40, 5 8)40   D [9] 5)45, 9 5)45, 5 9)45

3

A [5] 3)15, 5 3)15, 3 5)15   B [3] 3)9, 3 3)9

C [2] 3)6, 2 3)6, 3 2)6   D [4] 3)12, 4 3)12, 3 4)12

4

A [▨] 9)12, 1 9)9   B [▨] 9)19, 2 9)18   C [▨] 9)25, 2 9)18

D [▨] 9)39, 4 9)36   E [▨] 9)28, 3 9)27   F [▨] 9)15, 1 9)9

Lesson 17 (continued)

5

A 3 (crossed), 4)7, 2 (crossed), 5
B 5 (crossed), 9)54, -45 (crossed), 9
C $\begin{array}{r}6\\3\overline{)20}\\-18\\\hline 2\end{array}$
D 9 (crossed), 5)50, -45 (crossed), 5

6

A Each day Maria drank 3 glasses of milk. Maria drank 15 glasses of milk.

B Each day Jessie bought 9 apples. Jessie bought 27 apples.

C Fred washed 9 floors every week. Fred washed floors for 18 weeks.

D Lucy used 5 pens in every class. Lucy used pens in 10 classes.

E Every month Joe watered his plants 5 times. He watered his plants 10 times.

7

3 3)9   6 5)30   2 3)6   8 5)40   7 5)35   4 3)12   5 3)15

8 5)40   2 3)6   7 5)35   4 3)12   6 5)30   2 3)6   7 5)35

6 5)30   5 3)15   3 3)9   8 5)40

8 Finish the problems. If you can't subtract, make the answer smaller. Then copy the problem and the new answer in the space next to the problem.

A 2 (crossed), 9)22, -27 | 2 9)22 … 4
B 4 5)24, -20, 4
C 2 (crossed), 5)13, -25 | 2 5)13, -10, 3
D 2 (crossed), 9)24, -36 | 2 9)24, -18, 6

Lesson 17 (continued)

9 Write the fact for each problem.

A The big number in a times problem is 9. One small number is 3. What's the third number?

$$\dfrac{3}{3\,\overline{)9}}$$

B One small number in a times problem is 9. The other small number is 2. What's the third number?

$$\begin{array}{r} 9 \\ \times\,2 \\ \hline 18 \end{array}$$

C One small number in a times problem is 5. The other small number is 4. What's the third number?

$$\begin{array}{r} 5 \\ \times\,4 \\ \hline 20 \end{array}$$

D The big number in a times problem is 20. One small number is 5. What's the third number?

$$\dfrac{4}{5\,\overline{)20}}$$

10

A $\begin{array}{r}3\\3\overline{)11}\\-9\\\hline 2\end{array}$ B $\begin{array}{r}3\\9\overline{)30}\\-27\\\hline 3\end{array}$ C $\begin{array}{r}4\\5\overline{)21}\\-20\\\hline 1\end{array}$ D $\begin{array}{r}1\\3\overline{)5}\\-3\\\hline 2\end{array}$ E $\begin{array}{r}3\\9\overline{)34}\\-27\\\hline 7\end{array}$

F $\begin{array}{r}4\\3\overline{)14}\\-12\\\hline 2\end{array}$ G $\begin{array}{r}4\\5\overline{)20}\end{array}$ H $\begin{array}{r}2\\3\overline{)8}\\-6\\\hline 2\end{array}$ I $\begin{array}{r}2\\9\overline{)20}\\-18\\\hline 2\end{array}$ J $\begin{array}{r}4\\3\overline{)12}\end{array}$

K $\begin{array}{r}4\\9\overline{)38}\\-36\\\hline 2\end{array}$ L $\begin{array}{r}5\\3\overline{)16}\\-15\\\hline 1\end{array}$ M $\begin{array}{r}2\\5\overline{)11}\\-10\\\hline 1\end{array}$ N $\begin{array}{r}4\\3\overline{)13}\\-12\\\hline 1\end{array}$ O $\begin{array}{r}3\\9\overline{)29}\\-27\\\hline 2\end{array}$

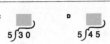

Lesson 18

Facts + Problems + Bonus = TOTAL

1

A. 5⟌35 — B. 5⟌40 — C. 5⟌50 — D. 5⟌45

2

A. 5⟌30 = [6]; 5⟌30 = 6; 6⟌30 = 5
B. 5⟌35 = [7]; 5⟌35 = 7; 7⟌35 = 5
C. 5⟌40 = [8]; 5⟌40 = 8; 8⟌40 = 5
D. 5⟌45 = [9]; 5⟌45 = 9; 9⟌45 = 5

3

A. 3⟌12 = 4 — B. 3⟌6 = 2 — C. 3⟌9 = 3 — D. 3⟌15 = 5

4

A. 8⟌48 = 5; −40; 8 (crossed out)
B. 4⟌29 = 6; −24; 5 (crossed out)
C. 3⟌16 = 4; −12; 4 (crossed out)
D. 7⟌49 = 6; −42; 7 (crossed out)
E. 9⟌32 = 3; −27; 5
F. 5⟌30 = 6; −30; 0

5

A. A dog chewed 6 bones every day. It chewed bones for 90 days. How many bones did the dog chew?

B. George cleaned 9 erasers each week. He cleaned 45 erasers. How many weeks did George clean erasers?

C. Marcy washed 3 cars every hour. She washed cars for 60 hours. How many cars did Marcy wash?

Lesson 18 (continued)

6

9⟌27 = 3; 5⟌35 = 7; 3⟌12 = 4; 5⟌30 = 6; 5⟌40 = 8; 9⟌36 = 4; 5⟌40 = 8

5⟌30 = 6; 3⟌6 = 2; 5⟌35 = 7; 3⟌9 = 3; 5⟌40 = 8; 5⟌30 = 6; 5⟌35 = 7

7

Write the facts with no remainders.

A. 9⟌40; 9⟌36 = 4
B. 9⟌47; 9⟌45 = 5
C. 9⟌20; 9⟌18 = 2
D. 9⟌16; 9⟌9 = 1
E. 9⟌6; 9⟌0 = 0
F. 9⟌43; 9⟌36 = 4

8

Finish the problems. If you can't subtract, make the answer smaller. Then copy the problem and the new answer in the space next to the problem.

A. 5⟌12 = 2; −10; 2
B. 9⟌17 = 2 (crossed out) 1; −18 — 9⟌17 = 1; −9; 8
C. 5⟌4 = 1 (crossed out) 0; −5 — 5⟌4 = 0; −0; 4
D. 9⟌46 = 5; −45; 1

9

A. 9⟌20 = 2; −18; 2
B. 5⟌17 = 3; −15; 2
C. 3⟌8 = 2; −6; 2
D. 3⟌10 = 3; −9; 1
E. 9⟌31 = 3; −27; 4

Lesson 19

Facts + Problems + Bonus = TOTAL

1

A. 5⟌40 = [8]; 5⟌40 = 8; 8⟌40 = 5
B. 5⟌35 = [7]; 5⟌35 = 7; 7⟌35 = 5
C. 5⟌50 = [10]; 5⟌50 = 10; 10⟌50 = 5
D. 5⟌30 = [6]; 5⟌30 = 6; 6⟌30 = 5

2

A. 4⟌21 = 6 (crossed out) 5; −16; 5
B. 7⟌32 = 4; −28; 4
C. 6⟌50 = 7 (crossed out) 8; −42; 8
D. 3⟌19 = 5 (crossed out) 6; −15; 4

3

A. Shu fixed 4 tires every day. He fixed tires for 32 days. How many tires did Shu fix?

B. Mr. Silbert wrote 2 stories every week. He wrote 12 stories. How many weeks did he write stories?

C. Carmen painted 5 cars every week. She painted 10 cars. How many weeks did Carmen paint?

4

A. Cora put 7 books in each pile. She used 21 books. How many piles did Cora make?

B. Larry baked 5 cakes each week. He baked cakes for 75 weeks. How many cakes did Larry bake?

C. Ms. Hightower ate 5 oranges each week. She ate oranges for 10 weeks. How many oranges did Ms. Hightower eat?

D. The Big Steel Company made 2 sheets of steel every hour. They made sheets of steel for 40 hours. How many sheets of steel did they make?

E. Mike picked 3 flowers every afternoon. He ended up with 15 flowers. How many afternoons did Mike pick flowers?

Lesson 19 (continued)

5

5⟌35 = 7; 5⟌45 = 9; 5⟌20 = 4; 5⟌40 = 8; 5⟌30 = 6; 5⟌15 = 3; 5⟌50 = 10

3⟌6 = 2; 9⟌18 = 2; 5⟌45 = 9; 9⟌36 = 4; 5⟌50 = 10; 5⟌25 = 5; 5⟌45 = 9

5⟌30 = 6; 3⟌15 = 5; 5⟌40 = 8; 5⟌50 = 10; 5⟌35 = 7

6

Write the facts with no remainders.

A. 9⟌23; 9⟌18 = 2
B. 9⟌17; 9⟌9 = 1
C. 9⟌42; 9⟌36 = 4
D. 9⟌12; 9⟌9 = 1
E. 9⟌37; 9⟌36 = 4
F. 9⟌20; 9⟌18 = 2
G. 3⟌5; 3⟌3 = 1
H. 3⟌10; 3⟌9 = 3
I. 3⟌13; 3⟌12 = 4
J. 3⟌14; 3⟌12 = 4
K. 3⟌17; 3⟌15 = 5
L. 3⟌8; 3⟌6 = 2

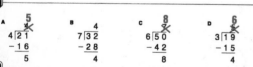

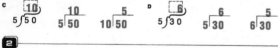

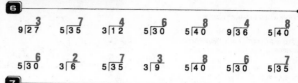

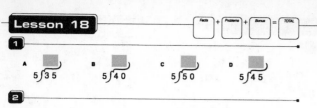

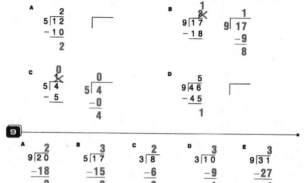

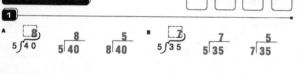

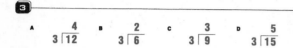

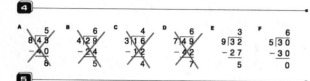

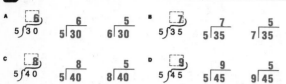

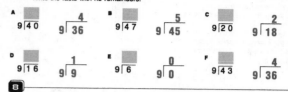

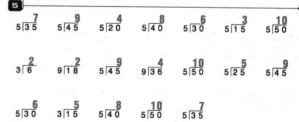

Write the fact for each problem.

A
The big number in a times problem is 9. One small number is 3. What is the third number?

$$3\overline{)9}$$ → 3

B
The big number in a times problem is 10. One small number is 5. What is the third number?

$$5\overline{)10}$$ → 2

C
One small number in a times problem is 10. The other small number is 5. What is the third number?

$$\begin{array}{r} 10 \\ \times\ 5 \\ \hline 50 \end{array}$$

D
One small number in a times problem is 6. The other small number is 4. What is the third number?

$$\begin{array}{r} 6 \\ \times\ 4 \\ \hline 24 \end{array}$$

E
The big number in a times problem is 15. One small number is 3. What is the third number?

$$3\overline{)15}$$ → 5

F
One small number in a times problem is 5. The other small number is 3. What is the third number?

$$\begin{array}{r} 5 \\ \times\ 3 \\ \hline 15 \end{array}$$

8

A
$$\begin{array}{r} 3 \\ 3\overline{)11} \\ -9 \\ \hline 2 \end{array}$$

B
$$\begin{array}{r} 3 \\ 5\overline{)18} \\ -15 \\ \hline 3 \end{array}$$

C
$$\begin{array}{r} 4 \\ 3\overline{)14} \\ -12 \\ \hline 2 \end{array}$$

D
$$\begin{array}{r} 3 \\ 3\overline{)10} \\ -9 \\ \hline 1 \end{array}$$

E
$$\begin{array}{r} 3 \\ 9\overline{)31} \\ -27 \\ \hline 4 \end{array}$$

Test + Facts + Problems + Bonus = TOTAL

1

A
$$5\overline{)40}\ \to 8 \qquad 5\overline{)40}\ \to 8 \qquad 8\overline{)40}\ \to 5$$

B
$$5\overline{)50}\ \to 10 \qquad 5\overline{)50}\ \to 10 \qquad 10\overline{)50}\ \to 5$$

C
$$5\overline{)45}\ \to 9 \qquad 5\overline{)45}\ \to 9 \qquad 9\overline{)45}\ \to 5$$

D
$$5\overline{)35}\ \to 7 \qquad 5\overline{)35}\ \to 7 \qquad 7\overline{)35}\ \to 5$$

2

A
$$\begin{array}{r} 8 \\ 8\overline{)66} \\ -56 \\ \hline 10 \end{array}$$

B
$$\begin{array}{r} 7 \\ 4\overline{)31} \\ -24 \\ \hline 7 \end{array}$$

C
$$\begin{array}{r} 9 \\ 7\overline{)63} \\ -56 \\ \hline 7 \end{array}$$

3

A	B	C	D	E
$8\overline{)0}$	$9\overline{)0}$	$4\overline{)0}$	$6\overline{)0}$	$3\overline{)0}$

4

A Kathy cut 3 lawns every week. She cut 12 lawns. How many weeks did Kathy work?

__weeks__ $3\overline{)12}$

B Joy put 5 hooks on each shelf. She put hooks on 90 shelves. How many hooks did Joy use?

__hooks__ $\begin{array}{r} 90 \\ \times\ 5 \end{array}$

C Stan lost 3 golf balls every time he played golf. He lost 15 golf balls. How many times did Stan play golf?

__times__ $3\overline{)15}$

D Lamar won 9 prizes every time he read a book. He read 18 books. How many prizes did Lamar win?

__prizes__ $\begin{array}{r} 18 \\ \times\ 9 \end{array}$

5

A
$$\begin{array}{r} 4 \\ 3\overline{)14} \\ -12 \\ \hline 2 \end{array}$$

B
$$\begin{array}{r} 3 \\ 9\overline{)30} \\ -27 \\ \hline 3 \end{array}$$

C
$$\begin{array}{r} 3 \\ 9\overline{)32} \\ -27 \\ \hline 5 \end{array}$$

D
$$\begin{array}{r} 1 \\ 5\overline{)9} \\ -5 \\ \hline 4 \end{array}$$

E
$$\begin{array}{r} 2 \\ 9\overline{)25} \\ -18 \\ \hline 7 \end{array}$$

6

$5\overline{)50}$ → 10 $9\overline{)27}$ → 3 $5\overline{)30}$ → 6 $5\overline{)45}$ → 9 $3\overline{)12}$ → 4 $5\overline{)40}$ → 8 $5\overline{)45}$ → 9

$3\overline{)9}$ → 3 $5\overline{)50}$ → 10 $9\overline{)45}$ → 5 $5\overline{)35}$ → 7 $5\overline{)20}$ → 4 $5\overline{)50}$ → 10 $5\overline{)15}$ → 3

$3\overline{)15}$ → 5 $5\overline{)45}$ → 9 $5\overline{)35}$ → 7 $5\overline{)30}$ → 6 $5\overline{)40}$ → 8 $5\overline{)10}$ → 2 $9\overline{)36}$ → 4

7

Write the facts with no remainders.

A
$9\overline{)25}$ $9\overline{)18}$ → 2

B
$9\overline{)3}$ $9\overline{)0}$ → 0

C
$9\overline{)14}$ $9\overline{)9}$ → 1

D
$9\overline{)41}$ $9\overline{)36}$ → 4

E
$9\overline{)16}$ $9\overline{)9}$ → 1

F
$9\overline{)21}$ $9\overline{)18}$ → 2

G
$3\overline{)7}$ $3\overline{)6}$ → 2

H
$3\overline{)16}$ $3\overline{)15}$ → 5

I
$3\overline{)1}$ $3\overline{)0}$ → 0

J
$3\overline{)11}$ $3\overline{)9}$ → 3

K
$3\overline{)4}$ $3\overline{)3}$ → 1

L
$3\overline{)13}$ $3\overline{)12}$ → 4

Facts + Problems + Bonus = TOTAL

1

A	B	C	D	E
$7\overline{)0}$	$2\overline{)0}$	$9\overline{)0}$	$5\overline{)0}$	$3\overline{)0}$

2

A $5\overline{)35}$ → 7 **B** $5\overline{)30}$ → 6 **C** $5\overline{)45}$ → 9 **D** $5\overline{)40}$ → 8

3

A
$$\begin{array}{r} 1 \\ 5\overline{)638} \\ -5 \\ \hline 1 \end{array}$$

B
$$\begin{array}{r} 9 \\ 5\overline{)4627} \\ -45 \\ \hline 1 \end{array}$$

C
$$\begin{array}{r} 1 \\ 5\overline{)532} \\ -5 \\ \hline 0 \end{array}$$

D
$$\begin{array}{r} 5 \\ 9\overline{)467} \\ -45 \\ \hline 1 \end{array}$$

4

A A student used 3 papers every lesson. She did 50 lessons. How many papers did the student use?

[150] __papers__ $\begin{array}{r} 50 \\ \times\ 3 \\ \hline 150 \end{array}$

B A seal balanced 5 balls in every show. It balanced 40 balls in shows. How many times was the seal in a show?

[8] __times__ $5\overline{)40}$ → 8

C Judy worked 3 hours every day. She worked 60 days. How many hours did Judy work?

[180] __hours__ $\begin{array}{r} 60 \\ \times\ 3 \\ \hline 180 \end{array}$

D Jane's store was open 5 hours every week. Her store was open for 30 hours. How many weeks was Jane's store open?

[6] __weeks__ $5\overline{)30}$ → 6

E Kim cut her hair 3 times every year. She cut her hair 90 times. How many years did Kim cut her hair?

[30] __years__ $3\overline{)90}$ → 30

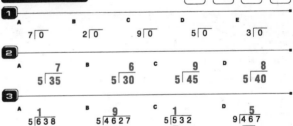

Lesson 21 (continued)

5

5⟌40 = 8	3⟌9 = 3	5⟌15 = 3	3⟌3 = 1	9⟌36 = 4	9⟌45 = 5	5⟌35 = 7
5⟌25 = 5	6⟌0 = 0	3⟌12 = 4	5⟌50 = 10	9⟌27 = 3	3⟌15 = 5	1⟌10 = 10
5⟌5 = 1	1⟌8 = 8	5⟌45 = 9	5⟌0 = 0	9⟌9 = 1	1⟌3 = 3	3⟌6 = 2
3⟌0 = 0	9⟌36 = 4	3⟌12 = 4	5⟌20 = 4	3⟌9 = 3	5⟌30 = 6	9⟌0 = 0

6

Write the facts with no remainders.

- A 3⟌7 = ☐ 3⟌6 = 2
- B 3⟌5 = ☐ 3⟌3 = 1
- C 3⟌10 = ☐ 3⟌9 = 3
- D 3⟌8 = ☐ 3⟌6 = 2
- E 3⟌16 = ☐ 3⟌15 = 5
- F 3⟌13 = ☐ 3⟌12 = 4

7

- A 5⟌33 = 6 (−30, 3)
- B 9⟌41 = 4 (−36, 5)
- C 3⟌14 = 4 (−12, 2)
- D 9⟌25 = 2 (−18, 7)
- E 5⟌6 = 1 (−5, 1)
- F 3⟌10 = 3 (−9, 1)
- G 9⟌21 = 2 (−18, 3)
- H 5⟌11 = 2 (−10, 1)
- I 9⟌38 = 4 (−36, 2)
- J 3⟌7 = 2 (−6, 1)
- K 5⟌44 = 8 (−40, 4)
- L 9⟌16 = 1 (−9, 7)

Lesson 22

Test + Facts + Problems + Bonus = TOTAL

1

- A 8⟌56 = ☐
- B 8⟌80 = ☐
- C 8⟌72 = ☐
- D 8⟌48 = ☐
- E 8⟌64 = ☐
- F 8⟌40 = ☐

2

- A 8⟌40 = [5]; 8⟌40 = 5; 5⟌40 = 8
- B 8⟌48 = [6]; 8⟌48 = 6; 6⟌48 = 8
- C 8⟌56 = [7]; 8⟌56 = 7; 7⟌56 = 8
- D 8⟌64 = [8]; 8⟌64 = 8
- E 8⟌72 = [9]; 8⟌72 = 9; 9⟌72 = 8
- F 8⟌80 = [10]; 8⟌80 = 10; 10⟌80 = 8

3

- A 5⟌473 = 9 (−45, 2)
- B 5⟌5621 = 1 (−5, 0)
- C 5⟌680 = 1 (−5, 1)
- D 3⟌162 = 5 (−15, 1)

4

A Hank painted 3 houses every week. He painted 15 houses. How many weeks did Hank paint houses?

3⟌15 = 5

[5] weeks

B Sally watches 5 trains each day. She watched 30 trains. How many days did Sally watch trains?

5⟌30 = 6

[6] days

Part 4 continues on the next page.

Lesson 22 (continued)

C Rico took 8 piano lessons every month. He took them for 7 months. How many lessons did Rico take?

8 × 7 = 56

[56] lessons

D Peg did 3 cartwheels for each cartoon show she watched on television. She did 12 cartwheels. How many cartoons did Peg watch?

3⟌12 = 4

[4] cartoons

E A plane carried 3 people on each trip. It made 6 trips. How many people rode in the plane?

3 × 6 = 18

[18] people

F Tony lost 9 pencils every year. He lost pencils for 3 years. How many pencils did Tony lose?

9 × 3 = 27

[27] pencils

5

9⟌36 = 4	5⟌15 = 3	3⟌3 = 1	5⟌35 = 7	9⟌27 = 3	3⟌9 = 3	5⟌50 = 10
7⟌0 = 0	5⟌30 = 6	5⟌25 = 5	3⟌15 = 5	2⟌0 = 0	9⟌45 = 5	5⟌5 = 1
5⟌40 = 8	5⟌20 = 4	3⟌6 = 2	9⟌18 = 2	5⟌10 = 2	5⟌40 = 8	3⟌12 = 4

6

8⟌56 = 7	8⟌40 = 5	5⟌35 = 7	3⟌6 = 2	8⟌48 = 6	5⟌40 = 8	5⟌30 = 6
5⟌50 = 10	8⟌56 = 7	5⟌45 = 9	8⟌40 = 5	3⟌15 = 5	8⟌48 = 6	8⟌40 = 5

Lesson 22 (continued)

7

Write the facts with no remainders.

- A 5⟌47 = ☐ 5⟌45 = 9
- B 5⟌36 = ☐ 5⟌35 = 7
- C 5⟌43 = ☐ 5⟌40 = 8
- D 5⟌42 = ☐ 5⟌40 = 8
- E 5⟌34 = ☐ 5⟌30 = 6
- F 5⟌46 = ☐ 5⟌45 = 9
- G 5⟌39 = ☐ 5⟌35 = 7
- H 5⟌31 = ☐ 5⟌30 = 6
- I 5⟌44 = ☐ 5⟌40 = 8

8

- A 3⟌11 = 3 (−9, 2)
- B 9⟌36 = 4
- C 9⟌31 = 3 (−27, 4)
- D 5⟌5 = 1
- E 9⟌20 = 2 (−18, 2)
- F 3⟌4 = 1 (−3, 1)
- G 5⟌13 = 2 (−10, 3)
- H 3⟌6 = 2
- I 9⟌47 = 5 (−45, 2)
- J 3⟌17 = 5 (−15, 2)
- K 5⟌18 = 3 (−15, 3)
- L 3⟌8 = 2 (−6, 2)
- M 9⟌31 = 3 (−27, 4)
- N 3⟌14 = 4 (−12, 2)
- O 5⟌27 = 5 (−25, 2)

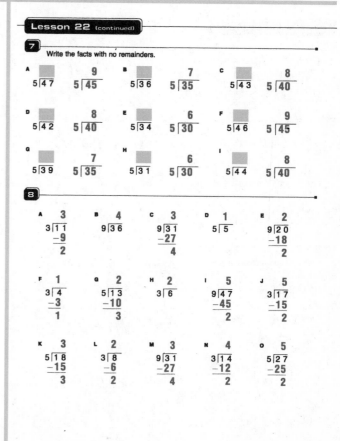

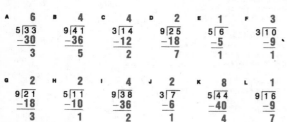

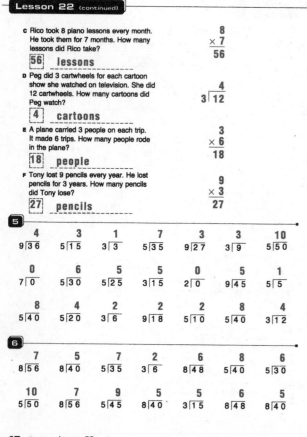

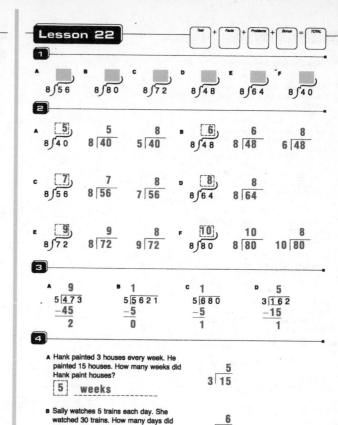

Lesson 23

Facts + Problems + Bonus = TOTAL

1

A	B	C	D	E	F
8)48	8)40	8)56	8)80	8)72	8)64

2

- **A** [5] 8)40 — 8)40 = 5 — 5)40 = 8
- **B** [6] 8)48 — 8)48 = 6 — 6)48 = 8
- **C** [7] 8)56 — 8)56 = 7 — 7)56 = 8
- **D** [8] 8)64 — 8)64 = 8
- **E** [9] 8)72 — 8)72 = 9 — 9)72 = 8
- **F** [10] 8)80 — 8)80 = 10 — 10)80 = 8

3

	A	B	C	D	E	F
Quotient	4	1	2	1	4	4
Problem	3)134	5)76	3)673	9)942	5)214	9)37
Subtract	−12	−5	−6	−9	−20	−36
Remainder	1	2	0	0	1	1

4

- **A** 9)379 = 42; −36↓; 19; −18; 1
- **B** 5)75 = 15; −5; 25; −25; 0
- **C** 3)124 = 41; −12; 04; −3; 1

Lesson 23 (continued)

5

A Each flower has 5 petals. There are 20 petals on the flowers. How many flowers are there?

5)20 = 4

[4] flowers

B Each motorcycle driver owns 6 helmets. There are 12 motorcycle drivers. How many helmets do they own?

12 × 6 = 72

[72] helmets

C There are 5 desks in each row. There are 15 rows. How many desks are there?

15 × 5 = 75

[75] desks

D Tam ate 3 meals every day. She ate 9 meals. How many days did Tam eat meals?

3)9 = 3

[3] days

E Nora was in 9 plays every month. She was in plays for 10 months. How many plays was Nora in?

10 × 9 = 90

[90] plays

F The president shook hands 9 times every minute. She shook hands 36 times. How many minutes did the president shake hands?

9)36 = 4

[4] minutes

G Rita read 5 comic books every hour. She read comics for 5 hours. How many comic books did Rita read?

5 × 5 = 25

[25] comic books

H Every time Lou went shopping, he went to 6 stores. He went shopping 9 times. How many stores did Lou go to?

9 × 6 = 54

[54] stores

Part 5 continues on the next page.

Lesson 23 (continued)

I Rosa told 3 jokes every time she went to a party. She told 9 jokes at parties. How many times did she go to a party?

3)9 = 3

[3] times

6

5)30 = 6	9)27 = 3	3)12 = 4	1)7 = 7	5)20 = 4	5)35 = 7	5)45 = 9
4)0 = 0	3)6 = 2	5)10 = 2	10)0 = 0	5)35 = 7	3)15 = 5	9)36 = 4
1)1 = 1	5)25 = 5	9)18 = 2	8)0 = 0	3)9 = 3	5)15 = 3	1)2 = 2
5)40 = 8	9)45 = 5	5)50 = 10	5)45 = 9	9)36 = 4	5)30 = 6	8)0 = 0

7

Write the facts with no remainders.

- **A** 5)38 — 5)35 = 7
- **B** 5)46 — 5)45 = 9
- **C** 5)33 — 5)30 = 6
- **D** 5)43 — 5)40 = 8
- **E** 5)31 — 5)30 = 6
- **F** 5)42 — 5)40 = 8
- **G** 5)39 — 5)35 = 7
- **H** 5)17 — 5)15 = 3
- **I** 5)48 — 5)45 = 9

Lesson 24

Facts + Problems + Bonus = TOTAL

1

- **A** [9] 8)72 — 8)72 = 9 — 9)72 = 8
- **B** [8] 8)64 — 8)64 = 8
- **C** [10] 8)80 — 8)80 = 10 — 10)80 = 8
- **D** [6] 8)48 — 8)48 = 6 — 6)48 = 8
- **E** [5] 8)40 — 8)40 = 5 — 5)40 = 8
- **F** [7] 8)56 — 8)56 = 7 — 7)56 = 8

2

	A	B	C	D	E	F
Quotient	3	9	3	1	5	7
Problem	9)318	5)482	3)102	9)940	3)172	5)391
Subtract	−27	−45	−9	−9	−15	−35
Remainder	4	3	1	0	2	4

3

- **A** 5)487 = 97; −45; 37; −35; 2
- **B** 5)298 = 59; −25; 48; −45; 3
- **C** 9)279 = 31; −27; 09; −9; 0
- **D** 3)74 = 24; −6; 14; −12; 2
- **E** 9)99 = 11; −9; 09; −9; 0

4

A Linda put together 9 bikes every day. She put together 27 bikes. How many days did Linda put bikes together?
9)27 = 3
[3] days

B There are 5 people on every team. There are 40 people. How many teams are there?
5)40 = 8
[8] teams

C There are 2 socks in every drawer. There are 4 drawers of socks. How many socks are there?
4 × 2 = 8
[8] socks

D The quarterback threw 3 passes every game. He threw 9 passes during games. How many games did he throw passes in?
3)9 = 3
[3] games

E Jerry saw 8 movies every week. He saw 72 movies. How many weeks did Jerry go to the movies?
8)72 = 9
[9] weeks

F There were 8 different bands in each parade. There were 4 parades. How many bands were in the parades?
8 × 4 = 32
[32] bands

G Each TV set has 5 knobs. There are 35 knobs on TV sets. How many TV sets are there?
5)35 = 7
[7] TV sets

H Mark sees 9 plants in every window. He sees 18 plants in windows. How many windows are there?
9)18 = 2
[2] windows

5

5)15 = 3	3)6 = 2	9)18 = 2	5)45 = 9	1)10 = 10	3)15 = 5	5)25 = 5
1)0 = 0	5)40 = 8	3)9 = 3	1)5 = 5	5)10 = 2	3)3 = 1	9)45 = 5
5)50 = 10	6)0 = 0	5)20 = 4	9)9 = 1	3)12 = 4	5)35 = 7	9)27 = 3
9)36 = 4	5)30 = 6	5)5 = 1	3)6 = 2	9)45 = 5	3)0 = 0	1)8 = 8

6

8)48 = 6	8)80 = 10	8)64 = 8	8)56 = 7	5)30 = 6	8)72 = 9	8)80 = 10
8)64 = 8	8)40 = 5	3)15 = 5	8)72 = 9	5)50 = 10	8)56 = 7	8)72 = 9
3)6 = 2	5)45 = 9	3)12 = 4	8)80 = 10	8)40 = 5	8)64 = 8	8)48 = 6

7

Write the facts with no remainders.

A 5)32 → 5)30 = 6
B 5)41 → 5)40 = 8
C 5)38 → 5)35 = 7
D 5)48 → 5)45 = 9
E 5)43 → 5)40 = 8
F 5)36 → 5)35 = 7

Lesson 25

Facts + Problems + Bonus = TOTAL

1

A [8] 8)64; 8)64 = 8
B [10] 8)80; 8)80 = 10; 10)80 = 8
C [5] 8)40; 8)40 = 5; 5)40 = 8
D [7] 8)56; 8)56 = 7; 7)56 = 8
E [9] 8)72; 8)72 = 9; 9)72 = 8
F [6] 8)48; 8)48 = 6; 6)48 = 8

2

A 5)82 = 16
−5
32
−30
2

B 3)45 = 15
−3
15
−15
0

3

A Karen brushed her teeth 3 times every day. She brushed her teeth for 6 days. How many times did Karen brush her teeth?
6 × 3 = 18
[18] times

B Each pie is cut into 8 pieces. There are 64 pieces of pie. How many pies are there?
8)64 = 8
[8] pies

C Harjo's Restaurant used 5 boxes of eggs every morning. The restaurant used 15 boxes of eggs. How many mornings did Harjo's Restaurant use eggs?
5)15 = 3
[3] mornings

Part 3 continues on the next page.

D There were 3 chairs at each table. There are 9 tables. How many chairs are there?
9 × 3 = 27
[27] chairs

E A horse ate 6 pails of oats each day. It ate oats for 12 days. How many pails of oats did the horse eat?
12 × 6 = 72
[72] pails

F Ken drove 8 kilometers every time he went to work. He drove 48 kilometers to work. How many times did Ken drive to work?
8)48 = 6
[6] times

G Kevin cooked 3 hot dogs for each person at the party. He cooked 12 hot dogs. How many people were at the party?
3)12 = 4
[4] people

4

5)35 = 7	9)18 = 2	3)0 = 0	5)15 = 3	9)36 = 4	3)9 = 3	3)0 = 0
5)25 = 5	5)0 = 0	5)40 = 8	3)12 = 4	9)0 = 0	1)8 = 8	5)20 = 4
3)6 = 2	5)50 = 10	9)27 = 3	1)6 = 6	5)10 = 2	9)0 = 0	5)45 = 9
3)15 = 5	5)30 = 6	9)45 = 5	5)50 = 10	9)27 = 3	3)12 = 4	5)0 = 0

Lesson 25 (continued)

5

8)64 = 8	3)9 = 3	8)40 = 5	8)72 = 9	8)56 = 7	8)80 = 10	8)48 = 6
8)72 = 9	5)35 = 7	8)64 = 8	5)50 = 10	8)56 = 7	8)80 = 10	5)45 = 9
5)30 = 6	3)12 = 4	8)80 = 10	3)6 = 2	8)72 = 9	8)40 = 5	8)64 = 8

6

- **A** 9)35 = 3; −27; 8
- **B** 3)14 = 4; −12; 2
- **C** 9)40 = 4; −36; 4
- **D** 5)17 = 3; −15; 2
- **E** 3)11 = 3; −9; 2

7

Work only the first part of these problems. Underline the first part. Then find the answer and the remainder for the first part.

- **A** 3)524 → 1; −3; 2
- **B** 5)135 → 2; −10; 3
- **C** 3)245 → 8; −24; 0
- **D** 5)823 → 1; −5; 3
- **E** 9)342 → 3; −27; 7

Lesson 26

Facts + Problems + Bonus = TOTAL

1

- **A** 8)56 = 7
- **B** 8)72 = 9
- **C** 8)64 = 8
- **D** 8)48 = 6

2

(50) 51 52 53 54 <u>55</u> 56 57 58 59 (60)

- **A** 58 rounds to 6 tens.
- **B** 53 rounds to 5 tens.
- **C** 56 rounds to 6 tens.
- **D** 57 rounds to 6 tens.
- **E** 51 rounds to 5 tens.
- **F** 54 rounds to 5 tens.

3

- **A** 9)138 = 15; −9; 48; −45; 3
- **B** 3)75 = 25; −6; 15; −15; 0
- **C** 5)119 = 23; −10; 19; −15; 4
- **D** 3)162 = 54; −15; 12; −12; 0
- **E** 5)258 = 51; −25; 08; −5; 3
- **F** 3)83 = 27; −6; 23; −21; 2

4

A Every classroom has 9 windows. There are 36 windows. How many classrooms are there?

[4] classrooms — 9)36 = 4

B Every day Jackie loaded 5 trucks. She loaded trucks for 8 days. How many trucks did Jackie load?

[40] trucks — 8 × 5 = 40

C Joan used 8 pieces of cheese each time she made pizza. She used 64 pieces of cheese. How many pizzas did Joan make?

[8] pizzas — 8)64 = 8

Part 4 continues on the next page.

Lesson 26 (continued)

D Each notebook has 32 pieces of paper. There are 8 notebooks. How many pieces of paper are there in all?

[256] pieces — 32 × 8 = 256

5

5)35 = 7	3)12 = 4	8)48 = 6	9)27 = 3	5)45 = 9	8)72 = 9	5)25 = 5
8)56 = 7	8)40 = 5	9)0 = 0	5)5 = 1	3)6 = 2	5)50 = 10	9)36 = 4
5)50 = 10	5)45 = 9	3)9 = 3	8)64 = 8	8)48 = 6	5)30 = 6	8)72 = 9
9)9 = 1	5)40 = 8	8)80 = 10	9)45 = 5	5)20 = 4	5)15 = 3	5)0 = 0
8)80 = 10	5)10 = 2	8)64 = 8	5)40 = 8	8)56 = 7	3)0 = 0	8)40 = 5
3)15 = 5	5)30 = 6	3)3 = 1	9)18 = 2	5)35 = 7		

6

Work only the first part of these problems. Find the first remainder and then stop.

- **A** 9)53 → 5; −45; 8
- **B** 5)278 → 5; −25; 2
- **C** 3)78 → 2; −6; 1
- **D** 9)402 → 4; −36; 4
- **E** 9)726 → 8; −72; 0

Lesson 27

Test + Facts + Problems + Bonus = TOTAL

1

- **A** 2)10
- **B** 2)6
- **C** 2)12
- **D** 2)4
- **E** 2)8
- **F** 2)2

2

- **A** 2)2 [1]; 2)2 = 1; 1)2 = 2
- **B** 2)4 [2]; 2)4 = 2
- **C** 2)6 [3]; 2)6 = 3; 3)6 = 2
- **D** 2)8 [4]; 2)8 = 4; 4)8 = 2
- **E** 2)10 [5]; 2)10 = 5; 5)10 = 2
- **F** 2)12 [6]; 2)12 = 6; 6)12 = 2

3

(30) 31 32 33 34 <u>35</u> 36 37 38 39 (40)

- **A** 36 rounds to 4 tens.
- **B** 33 rounds to 3 tens.
- **C** 34 rounds to 3 tens.
- **D** 32 rounds to 3 tens.
- **E** 37 rounds to 4 tens.
- **F** 39 rounds to 4 tens.

4

- **A** 5)2846 = 569; −25; 34; −30; 46; −45; 1
- **B** 9)4730 = 525; −45; 23; −18; 50; −45; 5
- **C** 3)7305 = 2435; −6; 13; −12; 10; −9; 15; −15; 0

Part 4 continues on the next page.

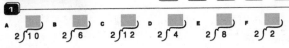

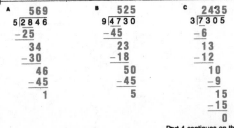

D

```
  421
9)3790
 -36
   19
  -18
   10
   -9
    1
```

E

```
  136
9)1231
 -9
  33
 -27
  61
 -54
   7
```

F

```
  434
5)2173
 -20
   17
  -15
   23
  -20
    3
```

5

A 3 girls can fit into every tent. 12 girls are going camping with tents. How many tents will they need?
[4] tents — 3)12 = 4

B Each branch has 8 leaves. There are 72 leaves. How many branches are there?
[9] branches — 8)72 = 9

C Each goal is worth 3 points. Fast Freddy scored 12 points. How many goals did he make?
[4] goals — 3)12 = 4

D Tim went to the grocery store 5 times every month. He went to the grocery store for 10 months. How many times did Tim go to the grocery store?
[50] times — 10 × 5 = 50

E Paula worked 5 days each week. She worked for 25 days. How many weeks did Paula work?
[5] weeks — 5)25 = 5

6

5)25 = 5	8)48 = 6	5)30 = 6	1)10 = 10	9)0 = 0	5)20 = 4	8)80 = 10
9)36 = 4	5)0 = 0	8)72 = 9	8)40 = 5	3)12 = 4	5)10 = 2	8)56 = 7
5)25 = 5	8)64 = 8	3)6 = 2	7)0 = 0	5)45 = 9	8)56 = 7	9)27 = 3
5)30 = 6	8)72 = 9	9)0 = 0	5)5 = 1	1)4 = 4	9)18 = 2	5)35 = 7
3)15 = 5	5)40 = 8	8)64 = 8	9)45 = 5	3)9 = 3	8)48 = 6	5)20 = 4
3)3 = 1	5)50 = 10	5)15 = 3	3)0 = 0	9)9 = 1		

7

2)6 = 3	8)56 = 7	2)8 = 4	2)4 = 2	8)40 = 5	2)6 = 3	8)72 = 9
2)4 = 2	8)64 = 8	8)80 = 10	2)8 = 4	8)48 = 6	2)4 = 2	2)6 = 3

8

Write the facts with no remainders.

A 8)45 — 8)40 = 5
B 8)75 — 8)72 = 9
C 8)86 — 8)80 = 10
D 8)67 — 8)64 = 8
E 8)70 — 8)64 = 8
F 8)52 — 8)48 = 6

Lesson 28

Facts + Problems + Bonus = TOTAL

1

A 2)8 **B** 2)12 **C** 2)6 **D** 2)10 **E** 2)4 **F** 2)2

2

A [1] 2)2 — 2)2 = 1 — 1)2 = 2
B [2] 2)4 — 2)4 = 2
C [3] 2)6 — 2)6 = 3 — 3)6 = 2
D [4] 2)8 — 2)8 = 4 — 4)8 = 2
E [5] 2)10 — 2)10 = 5 — 5)10 = 2
F [6] 2)12 — 2)12 = 6 — 6)12 = 2

3

A 36 rounds to 4 tens.
B 47 rounds to 5 tens.
C 93 rounds to 9 tens.
D 86 rounds to 9 tens.
E 11 rounds to 1 tens.
F 23 rounds to 2 tens.
G 78 rounds to 8 tens.
H 61 rounds to 6 tens.
I 52 rounds to 5 tens.
J 19 rounds to 2 tens.

4

A

```
  231
9)2081
 -18
   28
  -27
   11
   -9
    2
```

B

```
  867
8)6937
 -64
   53
  -48
   57
  -56
    1
```

C

```
  988
8)7904
 -72
   70
  -64
   64
  -64
    0
```

D

```
  519
5)2596
 -25
   09
   -5
   46
  -45
    1
```

E

```
  558
8)4470
 -40
   47
  -40
   70
  -64
    6
```


5

5)35 = 7	3)3 = 1	8)56 = 7	9)36 = 4	8)72 = 9	5)5 = 1	6)0 = 0
8)40 = 5	9)0 = 0	3)12 = 4	8)48 = 6	5)45 = 9	1)7 = 7	3)9 = 3
5)10 = 2	8)64 = 8	9)27 = 3	3)12 = 4	9)18 = 2	3)0 = 0	5)30 = 6
9)27 = 3	1)1 = 1	5)25 = 5	3)6 = 2	2)0 = 0	5)40 = 8	8)56 = 7
3)15 = 5	5)50 = 10	8)48 = 6	5)0 = 0	8)40 = 5	9)45 = 5	8)64 = 8
8)80 = 10	5)15 = 3	8)72 = 9	9)9 = 1	5)20 = 4		

6

8)48 = 6	2)8 = 4	2)4 = 2	8)40 = 5	2)6 = 3	8)64 = 8	2)6 = 3
8)56 = 7	2)8 = 4	8)72 = 9	2)4 = 2	8)80 = 10	2)6 = 3	2)8 = 4

7

Write the facts with no remainders.

A 8)77 — 8)72 = 9
B 8)60 — 8)56 = 7
C 8)44 — 8)40 = 5
D 8)83 — 8)80 = 10
E 8)53 — 8)48 = 6
F 8)59 — 8)56 = 7

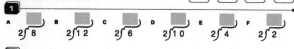

8

A
```
   3
3)11
 -9
  2
```
B
```
   5
8)45
 -40
   5
```
C
```
  11
3)35
 -3
  05
  -3
   2
```
D
```
   7
8)59
 -56
   3
```
E
```
   9
5)48
 -45
   3
```

9

A There are 2 paper clips in every box. There are 10 boxes. How many paper clips are there?

[20] paper clips

```
  10
× 2
  20
```

B Jill cooked dinner 8 times every month. She cooked dinner 40 times. How many months did Jill cook dinner?

[5] months

```
   5
8)40
```

C Each boat carries 5 people. There are 20 people in boats. How many boats are there?

[4] boats

```
   4
5)20
```


Facts + Problems + Bonus = TOTAL

1

A	2)6 → [3]	2)6 → 3	3)6 → 2	B	2)10 → [5]	2)10 → 5	5)10 → 2
C	2)4 → [2]	2)4 → 2		D	2)12 → [6]	2)12 → 6	6)12 → 2
E	2)2 → [1]	2)2 → 1	1)2 → 2	F	2)8 → [4]	2)8 → 4	4)8 → 2

2

A 82 rounds to 8 tens.
B 33 rounds to 3 tens.
C 44 rounds to 4 tens.
D 59 rounds to 6 tens.
E 97 rounds to 10 tens.
F 74 rounds to 7 tens.
G 26 rounds to 3 tens.
H 91 rounds to 9 tens.
I 12 rounds to 1 tens.
J 68 rounds to 7 tens.

3

A 64 rounds to 6 tens.
B 96 rounds to 10 tens.
C 84 rounds to 8 tens.
D 26 rounds to 3 tens.
E 24 rounds to 2 tens.
F 36 rounds to 4 tens.
G 14 rounds to 1 tens.
H 16 rounds to 2 tens.
I 54 rounds to 5 tens.
J 56 rounds to 6 tens.

4

A
```
    608
8)4867
 -48
   06
   -0
    67
   -64
     3
```
B
```
   102
5)514
 -5
  01
  -0
   14
  -10
    4
```
C
```
    504
3)1513
 -15
   01
   -0
    13
   -12
     1
```
D
```
   3045
9)9405
 -9
  04
  -0
   40
  -36
    45
   -45
     0
```
E
```
    709
8)5674
 -56
   07
  - 0
    74
   -72
     2
```

5

9)9 → 1	8)72 → 9	5)40 → 8	3)3 → 1	8)56 → 7	5)35 → 7	9)45 → 5
5)15 → 3	8)56 → 7	3)15 → 5	9)36 → 4	3)0 → 0	9)27 → 3	5)5 → 1
8)72 → 9	3)12 → 4	8)64 → 8	5)20 → 4	8)40 → 5	5)50 → 10	8)80 → 10
3)12 → 4	5)0 → 0	1)8 → 8	8)48 → 6	9)18 → 2	3)9 → 3	9)27 → 3
5)45 → 9	8)40 → 5	5)25 → 5	9)36 → 4	5)10 → 2	8)64 → 8	3)6 → 2
8)48 → 6	3)9 → 3	9)0 → 0	5)30 → 6	8)80 → 10		

6

2)8 → 4	2)2 → 1	2)10 → 5	2)4 → 2	8)64 → 8	2)12 → 6	2)10 → 5

Part 6 continues on the next page.

2)6 → 3	8)40 → 5	2)12 → 6	8)56 → 7	2)6 → 3	2)4 → 2	8)80 → 10
2)12 → 6	8)48 → 6	2)10 → 5	8)72 → 9	2)8 → 4	8)32 → 4	2)2 → 1

7

Write the facts with no remainders.

A	8)73	8)72 → 9	B	8)51	8)48 → 6
C	8)43	8)40 → 5	D	8)87	8)80 → 10
E	8)62	8)56 → 7	F	8)68	8)64 → 8

8

A There are 9 apples in every bag. There are 18 apples in bags. How many bags of apples are there?

[2] bags

```
   2
9)18
```

B Every hour Eric made 3 telephone calls. He made 15 calls. How many hours did Eric make calls?

[5] hours

```
   5
3)15
```

C Ming taught 4 classes every year. He taught for 4 years. How many classes did Ming teach?

[16] classes

```
   4
× 4
  16
```

D Sue went fishing 9 times every year. She went fishing 36 times. How many years did Sue go fishing?

[4] years

```
   4
9)36
```

Part 8 continues on the next page.

E. Marcy slept 4 hours each night. Eight nights went by. How many hours did Marcy sleep?

32 hours

$$\begin{array}{r} 4 \\ \times 8 \\ \hline 32 \end{array}$$

F. Each truck has 8 wheels. There are 48 wheels. How many trucks are there?

6 trucks

$$8\overline{)48} \quad 6$$

9

Work the problems and find the remainders.

A.
$$\begin{array}{r} 66 \\ 8\overline{)529} \\ -48 \\ \hline 49 \\ -48 \\ \hline 1 \end{array}$$

B.
$$\begin{array}{r} 246 \\ 3\overline{)740} \\ -6 \\ \hline 14 \\ -12 \\ \hline 20 \\ -18 \\ \hline 2 \end{array}$$

C.
$$\begin{array}{r} 45 \\ 9\overline{)407} \\ -36 \\ \hline 47 \\ -45 \\ \hline 2 \end{array}$$

D.
$$\begin{array}{r} 186 \\ 5\overline{)932} \\ -5 \\ \hline 43 \\ -40 \\ \hline 32 \\ -30 \\ \hline 2 \end{array}$$

E.
$$\begin{array}{r} 35 \\ 9\overline{)318} \\ -27 \\ \hline 48 \\ -45 \\ \hline 3 \end{array}$$

F.
$$\begin{array}{r} 96 \\ 5\overline{)482} \\ -45 \\ \hline 32 \\ -30 \\ \hline 2 \end{array}$$

G.
$$\begin{array}{r} 57 \\ 3\overline{)172} \\ -15 \\ \hline 22 \\ -21 \\ \hline 1 \end{array}$$

H.
$$\begin{array}{r} 78 \\ 5\overline{)391} \\ -35 \\ \hline 41 \\ -40 \\ \hline 1 \end{array}$$

1

A. $2\overline{)16}$ B. $2\overline{)20}$ C. $2\overline{)18}$ D. $2\overline{)14}$

2

A. $2\overline{)14} = 7$ $2\overline{)14} = 7$ $7\overline{)14} = 2$
B. $2\overline{)16} = 8$ $2\overline{)16} = 8$ $8\overline{)16} = 2$

C. $2\overline{)18} = 9$ $2\overline{)18} = 9$ $9\overline{)18} = 2$
D. $2\overline{)20} = 10$ $2\overline{)20} = 10$ $10\overline{)20} = 2$

3

A. $2\overline{)6} = 3$ B. $2\overline{)12} = 6$ C. $2\overline{)8} = 4$ D. $2\overline{)10} = 5$

4

A. 87 rounds to **9** tens.   B. 52 rounds to **5** tens.

C. 57 rounds to **6** tens.   D. 32 rounds to **3** tens.

E. 17 rounds to **2** tens.   F. 12 rounds to **1** tens.

G. 77 rounds to **8** tens.   H. 82 rounds to **8** tens.

5

A.
$$\begin{array}{r} 402 \\ 9\overline{)3624} \\ -36 \\ \hline 024 \\ -18 \\ \hline 6 \end{array}$$

B.
$$\begin{array}{r} 508 \\ 8\overline{)4065} \\ -40 \\ \hline 065 \\ -64 \\ \hline 1 \end{array}$$

C.
$$\begin{array}{r} 906 \\ 5\overline{)4530} \\ -45 \\ \hline 030 \\ -30 \\ \hline 0 \end{array}$$

D.
$$\begin{array}{r} 403 \\ 3\overline{)1210} \\ -12 \\ \hline 010 \\ -9 \\ \hline 1 \end{array}$$

E.
$$\begin{array}{r} 204 \\ 9\overline{)1838} \\ -18 \\ \hline 038 \\ -36 \\ \hline 2 \end{array}$$

6

A.
$$\begin{array}{r} 141 \\ 9\overline{)1270} \\ -9 \\ \hline 37 \\ -36 \\ \hline 10 \\ -9 \\ \hline 1 \end{array}$$

B.
$$\begin{array}{r} 958 \\ 8\overline{)7664} \\ -72 \\ \hline 46 \\ -40 \\ \hline 64 \\ -64 \\ \hline 0 \end{array}$$

C.
$$\begin{array}{r} 434 \\ 3\overline{)1304} \\ -12 \\ \hline 10 \\ -9 \\ \hline 14 \\ -12 \\ \hline 2 \end{array}$$

D.
$$\begin{array}{r} 1448 \\ 5\overline{)7240} \\ -5 \\ \hline 22 \\ -20 \\ \hline 24 \\ -20 \\ \hline 40 \\ -40 \\ \hline 0 \end{array}$$

E.
$$\begin{array}{r} 41 \\ 9\overline{)371} \\ -36 \\ \hline 11 \\ -9 \\ \hline 2 \end{array}$$

F.
$$\begin{array}{r} 558 \\ 8\overline{)4466} \\ -40 \\ \hline 46 \\ -40 \\ \hline 66 \\ -64 \\ \hline 2 \end{array}$$

7

$8\overline{)40}=5$ $5\overline{)15}=3$ $3\overline{)0}=0$ $2\overline{)10}=5$ $3\overline{)12}=4$ $9\overline{)27}=3$ $5\overline{)30}=6$

$3\overline{)6}=2$ $2\overline{)2}=1$ $9\overline{)45}=5$ $8\overline{)80}=10$ $5\overline{)10}=2$ $8\overline{)56}=7$ $5\overline{)45}=9$

$9\overline{)0}=0$ $5\overline{)35}=7$ $8\overline{)48}=6$ $2\overline{)8}=4$ $2\overline{)4}=2$ $5\overline{)20}=4$ $3\overline{)3}=1$

Part 7 continues on the next page.

$2\overline{)12}=6$ $3\overline{)9}=3$ $5\overline{)5}=1$ $9\overline{)18}=2$ $5\overline{)50}=10$ $2\overline{)0}=0$ $5\overline{)40}=8$

$9\overline{)9}=1$ $5\overline{)0}=0$ $2\overline{)8}=4$ $4\overline{)0}=0$ $8\overline{)72}=9$ $3\overline{)12}=4$ $5\overline{)25}=5$

$2\overline{)6}=3$ $3\overline{)15}=5$ $1\overline{)5}=5$ $8\overline{)64}=8$ $9\overline{)36}=4$

8

$2\overline{)10}=5$ $2\overline{)14}=7$ $2\overline{)12}=6$ $8\overline{)48}=6$ $2\overline{)16}=8$ $8\overline{)72}=9$ $2\overline{)6}=3$

$2\overline{)14}=7$ $2\overline{)16}=8$ $2\overline{)8}=4$ $8\overline{)56}=7$ $2\overline{)12}=6$ $2\overline{)16}=8$ $2\overline{)14}=7$

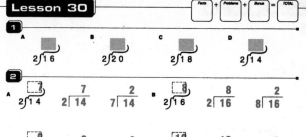

Lesson 31

Facts + Problems + Bonus = TOTAL

1

A 2⟌18 [ ] B 2⟌14 [ ] C 2⟌20 [ ] D 2⟌16 [ ]

2

A [7] 2⟌14; 2⟌14 = 7; 7⟌14 = 2 B [8] 2⟌16; 2⟌16 = 8; 8⟌16 = 2

C [9] 2⟌18; 2⟌18 = 9; 9⟌18 = 2 D [10] 2⟌20; 2⟌20 = 10; 10⟌20 = 2

3

A 3⟌61 = 20; −6; 01

B 9⟌97 = 10; −9; 07

C 3⟌92 = 30; −9; 02

D 5⟌157 = 31; −15; 07; −5; 2

E 8⟌86 = 10; −8; 06

F 9⟌182 = 20; −18; 02

4

3⟌15 = 5, 9⟌27 = 3, 2⟌4 = 2, 5⟌25 = 5, 2⟌8 = 4, 8⟌56 = 7, 3⟌6 = 2

5⟌40 = 8, 5⟌10 = 2, 8⟌48 = 6, 9⟌36 = 4, 3⟌9 = 3, 5⟌45 = 9, 2⟌10 = 5

3⟌12 = 4, 5⟌5 = 1, 9⟌45 = 5, 2⟌2 = 1, 5⟌20 = 4, 8⟌40 = 5, 9⟌18 = 2

3⟌3 = 1, 5⟌35 = 7, 8⟌72 = 9, 2⟌12 = 6, 8⟌56 = 7, 3⟌12 = 4, 5⟌50 = 10

Part 4 continues on the next page.

Lesson 31 (continued)

2⟌6 = 3, 5⟌30 = 6, 8⟌80 = 10, 5⟌35 = 7, 2⟌8 = 4, 5⟌15 = 3, 8⟌64 = 8

9⟌9 = 1, 8⟌72 = 9, 5⟌30 = 6, 2⟌12 = 6, 3⟌9 = 3

5

8⟌16 = 2, 2⟌14 = 7, 2⟌8 = 4, 2⟌12 = 6, 8⟌56 = 7, 2⟌16 = 8, 8⟌40 = 5

8⟌64 = 8, 2⟌10 = 5, 2⟌14 = 7, 8⟌48 = 6, 2⟌16 = 8, 8⟌72 = 9, 2⟌14 = 7

6

Write the facts with no remainders.

A [ ] 2⟌5; 2⟌4 = 2 B [ ] 2⟌13; 2⟌12 = 6 C [ ] 2⟌9; 2⟌8 = 4

D [ ] 2⟌7; 2⟌6 = 3 E [ ] 2⟌3; 2⟌2 = 1 F [ ] 2⟌11; 2⟌10 = 5

7

A 9⟌3720 = 413; −36; 12; −9; 30; −27; 3

B 3⟌613 = 204; −6; 013; −12; 1

C 5⟌8290 = 1658; −5; 32; −30; 29; −25; 40; −40; 0

D 8⟌4864 = 608; −48; 064; −64; 0

Lesson 31 (continued)

8

A Jack worked 8 hours every day. Jack worked for 48 hours. How many days did Jack work?

[6] days 8⟌48 = 6

B There are 2 liters of pop in each bottle. I have 10 bottles. How many liters of pop do I have?

[20] liters 10 × 2 = 20

C There are 2 flowers in each vase. There are 8 vases. How many flowers are there?

[16] flowers 8 × 2 = 16

D Ana skates 8 kilometers every day. How many days will it take her to skate 120 kilometers?

[15] days 8⟌120 = 15; −8; 40; −40; 0

Lesson 32

Test + Facts + Problems + Bonus = TOTAL

1

A [10] 2⟌20; 2⟌20 = 10; 10⟌20 = 2 B [7] 2⟌14; 2⟌14 = 7; 7⟌14 = 2

C [9] 2⟌18; 2⟌18 = 9; 9⟌18 = 2 D [8] 2⟌16; 2⟌16 = 8; 8⟌16 = 2

2

A 367 B 594 C 621 D 357

3

A 5⟌203 = 40; −20; 03

B 7⟌356 = 50; −35; 06

C 7⟌74 = 10; −7; 04

D 2⟌161 = 80; −16; 01

E 2⟌81 = 40; −8; 01

F 7⟌213 = 30; −21; 03

4

A 34⟌99 = 2; −68; 31

B 31⟌75 = 2; −62; 13

C 12⟌47 = 3; −36; 11

5

8⟌48 = 6, 5⟌0 = 0, 9⟌9 = 1, 2⟌4 = 2, 2⟌8 = 4, 5⟌25 = 5, 8⟌72 = 9

2⟌2 = 1, 5⟌10 = 2, 2⟌12 = 6, 8⟌56 = 7, 5⟌50 = 10, 1⟌9 = 9, 3⟌3 = 1

3⟌15 = 5, 9⟌45 = 5, 1⟌6 = 6, 5⟌5 = 1, 8⟌40 = 5, 5⟌35 = 7, 2⟌8 = 4

5⟌40 = 8, 2⟌12 = 6, 8⟌56 = 7, 3⟌12 = 4, 9⟌27 = 3, 2⟌10 = 5, 5⟌15 = 3

Part 5 continues on the next page.

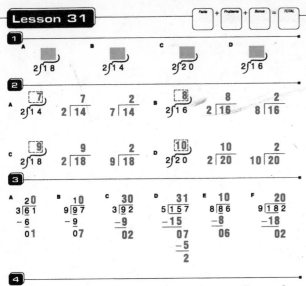

8⟌80 = 10	9⟌18 = 2	5⟌30 = 6	2⟌6 = 3	8⟌72 = 9	3⟌9 = 3	8⟌64 = 8
3⟌6 = 2	5⟌20 = 4	9⟌36 = 4	8⟌0 = 0	5⟌45 = 9		

6

2⟌16 = 8	2⟌10 = 5	2⟌18 = 9	2⟌8 = 4	2⟌20 = 10	2⟌14 = 7	2⟌18 = 9
8⟌64 = 8	2⟌20 = 10	2⟌16 = 8	2⟌14 = 7	2⟌18 = 9	2⟌6 = 3	2⟌20 = 10
8⟌48 = 6	2⟌4 = 2	2⟌12 = 6				

7

Write the facts with no remainders.

A 2⟌1 → 2⟌0 = 0 B 2⟌7 → 2⟌6 = 3 C 2⟌13 → 2⟌12 = 6

D 2⟌5 → 2⟌4 = 2 E 2⟌11 → 2⟌10 = 5 F 2⟌3 → 2⟌2 = 1

8

A Each bird has 4 worms. There are 12 birds. How many worms are there?
48 worms (12 × 4 = 48)

B There are 8 chairs for every table at the store. There are 48 chairs at the store. How many tables are there?
6 tables (8⟌48 = 6)

Part 8 continues on the next page.

C Leo has 9 records in each box. He has 27 records. How many boxes does he have?
3 boxes (9⟌27 = 3)

D A store sells 4 shirts every week. The store sells shirts for 56 weeks. How many shirts did the store sell?
224 shirts (56 × 4 = 224)

9

A 5⟌1525 = 305
−15, 025, −25, 0

B 2⟌815 = 407
−8, 015, −14, 1

C 5⟌8325 = 1665
−5, 33, −30, 32, −30, 25, −25, 0

D 9⟌931 = 103
−9, 031, −27, 4

E 2⟌1368 = 684
−12, 16, −16, 8, −8, 0

Lesson 33

Facts + Problems + Bonus = TOTAL

1

A 2⟌16 B 2⟌18 C 2⟌14 D 2⟌20

2

A 2⟌16: 8 — 2⟌16 = 8, 8⟌16 = 2 B 2⟌20: 10 — 2⟌20 = 10, 10⟌20 = 2

C 2⟌14: 7 — 2⟌14 = 7, 7⟌14 = 2 D 2⟌18: 9 — 2⟌18 = 9, 9⟌18 = 2

3

A 7⟌35: 5 — 7⟌35 = 5, 5⟌35 = 7 B 7⟌14: 2 — 7⟌14 = 2, 2⟌14 = 7

C 7⟌21: 3 — 7⟌21 = 3, 3⟌21 = 7 D 7⟌28: 4 — 7⟌28 = 4, 4⟌28 = 7

4

A 861 B 479 C 325 D 764

5

A 21⟌79 = 3, −63, 16
B 13⟌472 = 3, −39, 8
C 42⟌583 = 1, −42, 16
D 21⟌42 = 2, −42, 0
E 32⟌149 = 4, −128, 21
F 24⟌368 = 1, −24, 12
G 31⟌137 = 4, −124, 13
H 14⟌38 = 2, −28, 10

6

A Kathy drills 75 holes. She drills 5 holes every minute. How many minutes does it take her to drill the holes?
15 minutes (5⟌75 = 15; −5, 25, −25, 0)

B A teacher bought 8 bags. Each bag has 214 peanuts. How many peanuts did the teacher buy?
1712 peanuts (214 × 8 = 1712)

C First Street Beach is 8 blocks long. There are 624 people on the beach. How many people are on each block of the beach?
78 people (8⟌624 = 78; −56, 64, −64, 0)

D Shirley rode her bike in 8 races. Each race was 328 meters long. How many meters did Shirley ride her bike during the races?
2624 meters (328 × 8 = 2624)

7

8⟌56 = 7	5⟌20 = 4	2⟌6 = 3	9⟌27 = 3	2⟌4 = 2	5⟌5 = 1	8⟌40 = 5
5⟌45 = 9	1⟌3 = 3	9⟌18 = 2	2⟌2 = 1	5⟌50 = 10	3⟌9 = 3	3⟌0 = 0
8⟌48 = 6	2⟌12 = 6	5⟌15 = 3	9⟌9 = 1	8⟌64 = 8	2⟌4 = 2	5⟌35 = 7
9⟌0 = 0	2⟌10 = 5	8⟌72 = 9	5⟌40 = 8	1⟌10 = 10	3⟌15 = 5	2⟌8 = 4
5⟌25 = 5	3⟌3 = 1	5⟌20 = 4	2⟌12 = 6	8⟌80 = 10	9⟌36 = 4	5⟌10 = 2
3⟌6 = 2	5⟌30 = 6	3⟌12 = 4	9⟌45 = 5	5⟌25 = 5		

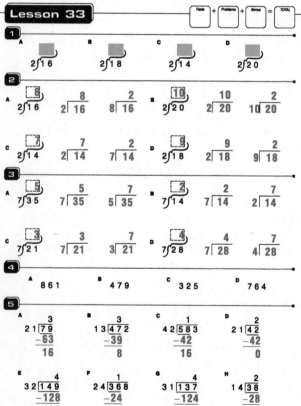

8

9 $2\overline{)18}$	3 $7\overline{)21}$	2 $7\overline{)14}$	9 $2\overline{)18}$	3 $7\overline{)21}$	10 $2\overline{)20}$	2 $7\overline{)14}$
7 $2\overline{)14}$	10 $2\overline{)20}$	8 $2\overline{)16}$	3 $7\overline{)21}$	10 $2\overline{)20}$	2 $7\overline{)14}$	9 $2\overline{)18}$

9

A
```
  605
2)1210
 -12
  010
  -10
    0
```

B
```
  70
8)562
 -56
  02
```

C
```
  403
9)3627
 -36
  027
  -27
    0
```

D
```
 20
3)61
 -6
 01
```

E
```
  3676
2)7352
 -6
  13
 -12
  15
 -14
  12
 -12
   0
```

F
```
  7
8)62
 -56
  6
```

G
```
  50
9)458
 -45
  08
```

H
```
  3
9)31
 -27
  4
```

I
```
  693
2)1387
 -12
  18
 -18
  07
  -6
   1
```

J
```
  80
8)642
 -64
  02
```

K
```
  51
8)408
 -40
  08
  -8
   0
```

L
```
  650
5)3254
 -30
  25
 -25
  04
```

1

A	B	C	D	E
$7\overline{)28}$	$7\overline{)14}$	$7\overline{)35}$	$7\overline{)21}$	$7\overline{)7}$

2

A $7\overline{)7}^{\boxed{1}}$ $7\overline{)7}^{1}$ $1\overline{)7}^{7}$ B $7\overline{)14}^{\boxed{2}}$ $7\overline{)14}^{2}$ $2\overline{)14}^{7}$

C $7\overline{)21}^{\boxed{3}}$ $7\overline{)21}^{3}$ $3\overline{)21}^{7}$ D $7\overline{)28}^{\boxed{4}}$ $7\overline{)28}^{4}$ $4\overline{)28}^{7}$

E $7\overline{)35}^{\boxed{5}}$ $7\overline{)35}^{5}$ $5\overline{)35}^{7}$

3

A	B	C	D
7 $2\overline{)14}$	10 $2\overline{)20}$	8 $2\overline{)16}$	9 $2\overline{)18}$

4

A	B	C	D
146	391	605	992

5

A 798 rounds to __80__ tens.   B 474 rounds to __47__ tens.

C 203 rounds to __20__ tens.   D 645 rounds to __65__ tens.

E 155 rounds to __16__ tens.   F 297 rounds to __30__ tens.

G 579 rounds to __58__ tens.   H 512 rounds to __51__ tens.

6

A
```
   3
32)985
  -96
    2
```

B
```
  1
7)83
 -73
  10
```

C
```
   6
10)664
  -60
    6
```

D
```
   4
41)197
  -164
    33
```

Part 6 continues on the next page.

E
```
    3
23)904
  -69
   21
```

F
```
    3
63)199
  -189
   10
```

G
```
    2
14)325
  -28
    4
```

7

4 $9\overline{)36}$	10 $2\overline{)20}$	7 $5\overline{)35}$	1 $3\overline{)3}$	2 $2\overline{)4}$	3 $5\overline{)15}$	10 $5\overline{)50}$
6 $8\overline{)48}$	2 $3\overline{)6}$	7 $8\overline{)56}$	7 $2\overline{)14}$	6 $5\overline{)30}$	10 $8\overline{)80}$	2 $9\overline{)18}$
5 $2\overline{)10}$	3 $3\overline{)9}$	5 $5\overline{)25}$	8 $2\overline{)16}$	3 $9\overline{)27}$	1 $2\overline{)2}$	9 $5\overline{)45}$
5 $8\overline{)40}$	4 $5\overline{)20}$	8 $2\overline{)16}$	6 $2\overline{)12}$	2 $2\overline{)4}$	8 $8\overline{)64}$	2 $5\overline{)10}$
6 $8\overline{)48}$	4 $2\overline{)8}$	4 $3\overline{)12}$	8 $5\overline{)40}$	1 $5\overline{)5}$	5 $3\overline{)15}$	3 $2\overline{)6}$
5 $9\overline{)45}$	9 $8\overline{)72}$	4 $2\overline{)8}$	9 $2\overline{)18}$	1 $9\overline{)9}$		

8

5 $2\overline{)10}$	2 $7\overline{)14}$	2 $2\overline{)4}$	9 $2\overline{)18}$	6 $2\overline{)12}$	4 $2\overline{)8}$	3 $7\overline{)21}$
3 $7\overline{)21}$	1 $2\overline{)2}$	2 $7\overline{)14}$	8 $2\overline{)16}$	10 $2\overline{)20}$	3 $7\overline{)21}$	3 $2\overline{)6}$

9

A
```
  3692
2)7385
 -6
  13
 -12
  18
 -18
   5
  -4
   1
```

B
```
  430
8)3447
 -32
  24
 -24
   7
```

C
```
  204
9)1843
 -18
  43
 -36
   7
```

D
```
  732
5)3661
 -35
  16
 -15
  11
 -10
   1
```

10

A There are 8 seats in each boat. There are 336 seats. How many boats are there?

[**42**]   **boats**

```
   42
8)336
 -32
  16
 -16
   0
```

B Randy corrected papers for 294 class periods. He corrected 7 papers every class period. How many papers did he correct?

[**2058**]   **papers**

```
  294
×   7
 2058
```

C Every morning Jason writes 3 pages in his book. He writes for 78 mornings. How many pages long will his book be?

[**234**]   **pages**

```
  78
×  3
 234
```

D Phil has 732 coins in a jar. Each day he puts 3 coins in the jar. How many days has Phil been saving coins?

[**244**]   **days**

```
  244
3)732
 -6
  13
 -12
  12
 -12
   0
```

Lesson 35

Test + Facts + Problems + Bonus = TOTAL

1

- A [3] 7)21   3: 7)21   7: 3)21
- B [4] 7)28   4: 7)28   7: 4)28
- C [1] 7)7   1: 7)7   7: 1)7
- D [5] 7)35   5: 7)35   7: 5)35
- E [2] 7)14   2: 7)14   7: 2)14

2

- A 265 rounds to 27 tens.
- B 304 rounds to 30 tens.
- C 737 rounds to 74 tens.
- D 972 rounds to 97 tens.
- E 180 rounds to 18 tens.
- F 846 rounds to 85 tens.
- G 591 rounds to 59 tens.
- H 459 rounds to 46 tens.

3

	A	B	C	D	E
Estimate / quotient	2 / 3	2 / 9	2 / 6	2 / 3	3 / 6
Problem	18)65	23)210	34)228	47)150	46)290
Subtract	−54	−207	−204	−141	−276
Remainder	11	3	24	9	14

4

- A 2)81 = 40: −8, 1
- B 2)7358 = 3679: −6, 13, −12, 15, −14, 18, −18, 0
- C 8)4018 = 502: −40, 18, −16, 2
- D 7)914 = 130: −7, 21, −21, 4
- E 5)3585 = 717: −35, 8, −5, 35, −35, 0

Lesson 35 (continued)

5

5)25 = 5	8)48 = 6	2)6 = 3	1)4 = 4	9)36 = 4	3)12 = 4	5)10 = 2
5)15 = 3	3)3 = 1	2)0 = 0	9)18 = 2	8)56 = 7	5)45 = 9	2)2 = 1
2)20 = 10	5)40 = 8	2)4 = 2	8)56 = 7	3)9 = 3	1)8 = 8	5)30 = 6
8)40 = 5	5)20 = 4	3)15 = 5	5)5 = 1	2)10 = 5	9)9 = 1	5)50 = 10
9)27 = 3	2)8 = 4	6)0 = 0	8)80 = 10	2)16 = 8	5)35 = 7	2)12 = 6
8)72 = 9	8)48 = 6	8)64 = 8	3)6 = 2	9)45 = 5		

6

2)10 = 5	2)18 = 9	7)28 = 4	7)14 = 2	7)35 = 5	2)12 = 6	7)35 = 5
7)28 = 4	2)16 = 8	7)21 = 3	2)14 = 7	7)35 = 5	7)28 = 4	2)6 = 3

7

Write the facts with no remainders.

- A 2)17   2)16 = 8
- B 2)19   2)18 = 9
- C 2)15   2)14 = 7
- D 2)11   2)10 = 5
- E 2)15   2)14 = 7
- F 2)17   2)16 = 8

Lesson 35 (continued)

8

A Ed fried 36 eggs. He fried 2 eggs every morning. How many mornings did Ed fry eggs?

18 mornings

2)36 = 18: −2, 16, −16, 0

B Every time Dave bought stamps he wrote 2 letters. He bought stamps 32 times. How many letters did Dave write?

64 letters

32 × 2 = 64

C There are 9 rows of plants. Each row has 173 plants. How many plants are there?

1557 plants

173 × 9 = 1557

D There are 9 students in the blue group. Each student in the blue group earned 387 points. How many points did the student earn in all?

3483 points

387 × 9 = 3483

9

Multiply to find the number you subtract. Then find the first remainder.

	A	B	C	D
Quotient	4	4	2	3
Problem	72)3092	90)400	43)891	13)507
Subtract	−288	−360	−86	−39
Remainder	21	40	3	11

Lesson 36

Facts + Problems + Bonus = TOTAL

1

- A 4)28
- B 4)40
- C 4)32
- D 4)36

2

- A [6] 4)24   6: 4)24   4: 6)24
- B [7] 4)28   7: 4)28   4: 7)28
- C [8] 4)32   8: 4)32   4: 8)32
- D [9] 4)36   9: 4)36   4: 9)36
- E [10] 4)40   10: 4)40   4: 10)40

3

	A	B	C	D	E
Estimate / quotient	3 / 6	1 / 2	2 / 5	1 / 3	4 / 6
Problem	16)971	37)958	54)300	46)1572	48)321
Subtract	−96	−74	−270	−138	−288
Remainder	1	21	30	19	33

4

- A [2] 7)14   2: 7)14   7: 2)14
- B [5] 7)35   5: 7)35   7: 5)35
- C [4] 7)28   4: 7)28   7: 4)28
- D [1] 7)7   1: 7)7   7: 1)7
- E [3] 7)21   3: 7)21   7: 3)21

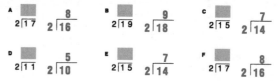

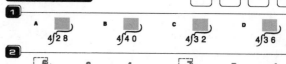

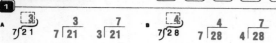

5

A 41)378 → 8
−328
50

B 18)72 → 3
−54
18

C 24)138 → 5
−120
18

D 46)233 → 4
−184
49

E 38)287 → 7
−266
21

6

2)8 → 4	3)3 → 1	8)40 → 5	5)50 → 10	9)18 → 2	2)4 → 2	5)30 → 6
8)72 → 9	2)16 → 8	5)20 → 4	2)2 → 1	3)9 → 3	8)48 → 6	9)36 → 4
5)45 → 9	2)10 → 5	8)64 → 8	2)8 → 4	5)40 → 8	5)15 → 3	2)6 → 3
5)25 → 5	3)12 → 4	9)27 → 3	3)6 → 2	2)14 → 7	7)0 → 0	5)10 → 2
2)20 → 10	1)2 → 2	2)12 → 6	5)35 → 7	9)45 → 5	8)56 → 7	3)15 → 5
2)16 → 8	2)12 → 6	8)80 → 10	9)9 → 1	2)18 → 9		

7

7)21 → 3	4)36 → 9	4)32 → 8	7)28 → 4	4)40 → 10	7)14 → 2	7)35 → 5
4)36 → 9	4)40 → 10	7)28 → 4	4)32 → 8	7)35 → 5	4)36 → 9	7)28 → 4

8

Write the facts with no remainders.

A 2)11 ▢ 2)10 → 5
B 2)15 ▢ 2)14 → 7
C 2)19 ▢ 2)18 → 9
D 2)17 ▢ 2)16 → 8
E 2)19 ▢ 2)18 → 9
F 2)15 ▢ 2)14 → 7

9

Multiply to find the number you subtract. Then find the first remainder.

A 10)825 → 8
−80
2

B 24)58 → 2
−48
10

C 84)198 → 2
−168
30

D 31)985 → 3
−93
5

10

A 7)728 → 104
−7
28
−28
0

B 2)1141 → 570
−10
14
−14
01

C 5)2765 → 553
−25
26
−25
15
−15
0

D 8)4056 → 507
−40
56
−56
0

11

A 407 rounds to 41 tens.
B 525 rounds to 53 tens.
C 649 rounds to 65 tens.
D 911 rounds to 91 tens.
E 397 rounds to 40 tens.

12

A Each team will have 9 children. There are 315 children who want to be on teams. How many teams will there be?
35 teams
9)315 → 35
−27
45
−45
0

B The ABC Company is going to make 96 trucks. Each truck will have 8 tires. How many tires will the ABC Company need?
768 tires
96 × 8 = 768

C There are 8 windows in each house. There are 124 houses. How many windows are there in all?
992 windows
124 × 8 = 992

D Harry drank 5 glasses of water each day. He drank 150 glasses of water. How many days did he drink water?
30 days
5)150 → 30

Lesson 37

Facts + Problems + Bonus = TOTAL

1

A 10)50 ▢
B 10)80 ▢
C 10)40 ▢
D 10)70 ▢

2

A [1] 10)10 10)10 → 1 1)10 → 10
B [2] 10)20 10)20 → 2 2)20 → 10
C [3] 10)30 10)30 → 3 3)30 → 10
D [4] 10)40 10)40 → 4 4)40 → 10
E [5] 10)50 10)50 → 5 5)50 → 10
F [6] 10)60 10)60 → 6 6)60 → 10
G [7] 10)70 10)70 → 7 7)70 → 10
H [8] 10)80 10)80 → 8 8)80 → 10
I [9] 10)90 10)90 → 9 9)90 → 10
J [10] 10)100 10)100 → 10

3

A 4)32 ▢
B 4)24 ▢
C 4)36 ▢
D 4)28 ▢

4

A [6] 4)24 4)24 → 6 6)24 → 4
B [7] 4)28 4)28 → 7 7)28 → 4
C [8] 4)32 4)32 → 8 8)32 → 4
D [9] 4)36 4)36 → 9 9)36 → 4
E [10] 4)40 4)40 → 10 10)40 → 4

5

A 14)98 → ~~7~~ 7
−84
14
14)98 → 7
−98
0

B 28)142 → ~~5~~ 5
−112
30
28)142 → 5
−140
2

C 33)191 → 5
−165
26

D 24)198 → ~~8~~ 8
−168
30
24)198 → 8
−192
6

6

2)18 → 9	8)48 → 6	5)35 → 7	7)14 → 2	9)45 → 5	3)12 → 4	2)8 → 4
7)21 → 3	5)5 → 1	2)16 → 8	9)36 → 4	8)56 → 7	2)2 → 1	5)40 → 8
8)40 → 5	2)10 → 5	5)20 → 4	3)9 → 3	7)28 → 4	5)50 → 10	2)14 → 7
5)45 → 9	7)7 → 1	7)21 → 3	2)4 → 2	5)30 → 6	8)64 → 8	5)15 → 3
2)12 → 6	5)25 → 5	3)6 → 2	8)72 → 9	2)20 → 10	9)27 → 3	2)6 → 3
5)10 → 2	7)35 → 5	8)80 → 10	9)18 → 2	3)15 → 5		

7

10)20 → 2	10)40 → 4	4)32 → 8	10)50 → 5	4)40 → 10	4)36 → 9	10)50 → 5
4)40 → 10	4)36 → 9	10)10 → 1	10)30 → 3	4)32 → 8	10)40 → 4	4)36 → 9

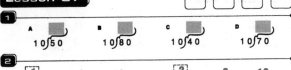

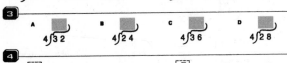

Lesson 37 (continued)

8 Write the facts with no remainders.

A [shaded box] 2|17 — 8: 2|16
B [shaded box] 2|15 — 7: 2|14
C [shaded box] 2|21 — 10: 2|20
D [shaded box] 2|19 — 9: 2|18
E [shaded box] 2|13 — 6: 2|12
F [shaded box] 2|11 — 5: 2|10

9 Multiply to find the number you subtract. Then find the first remainder.

A 3: 26|1014, −78, 23
B 4: 43|17,259, −172, 0
C 3: 32|114, −96, 18
D 2: 29|674, −58, 9
E 2: 14|3495, −28, 6

10

A 108 rounds to 11 tens.
B 929 rounds to 93 tens.
C 860 rounds to 86 tens.
D 666 rounds to 67 tens.
E 345 rounds to 35 tens.
F 378 rounds to 38 tens.

11

A A teacher came to school for 120 days. She drove 5 kilometers every day. How many kilometers did she drive?
120 × 5 = 600
600 kilometers

B 125 cars came into a parking lot. There were 5 people in each car. How many people came into the parking lot?
125 × 5 = 625
625 people

C A machine wrapped 9 loaves of bread every minute. It wrapped 1260 loaves of bread. How many minutes did it take to wrap the loaves?
140: 9|1260, −9, 36, −36, 0
140 minutes

D Ray uses 7 liters of paint each week. He has used 2128 liters. How many weeks has Ray painted?
304: 7|2128, −21, 28, −28, 0
304 weeks

Lesson 38

Facts + Problems + Bonus = TOTAL

1

A [shaded box] 10|30
B [shaded box] 10|60
C [shaded box] 10|90
D [shaded box] 10|100

2

A 1: 10|10 — 1: 10|10 — 10: 1|10
B 2: 10|20 — 2: 10|20 — 10: 2|20
C 3: 10|30 — 3: 10|30 — 10: 3|30
D 4: 10|40 — 4: 10|40 — 10: 4|40
E 5: 10|50 — 5: 10|50 — 10: 5|50
F 6: 10|60 — 6: 10|60 — 10: 6|60
G 7: 10|70 — 7: 10|70 — 10: 7|70
H 8: 10|80 — 8: 10|80 — 10: 8|80
I 9: 10|90 — 9: 10|90 — 10: 9|90
J 10: 10|100 — 10: 10|100

3

A 5 (crossed out): 31|169, −124, 45 — 5: 31|169, −155, 14
B 3 (crossed out): 14|42, −28, 14 — 3: 14|42, −42, 0
C 6: 23|157, −138, 19
D 4 (crossed out): 45|182, −135, 47 — 4: 45|182, −180, 2

Lesson 38 (continued)

4

A 7: 4|28 — 7: 4|28 — 4: 7|28
B 8: 4|32 — 8: 4|32 — 4: 8|32
C 10: 4|40 — 10: 4|40 — 4: 10|40
D 9: 4|36 — 9: 4|36 — 4: 9|36
E 6: 4|24 — 6: 4|24 — 4: 6|24

5

8: 2|16	5: 8|40	2: 2|4	5: 3|15	3: 9|27	2: 7|14	4: 5|20
10: 8|80	4: 3|12	9: 5|45	4: 7|28	7: 5|35	1: 2|2	7: 8|56
3: 2|6	1: 7|7	3: 5|15	7: 2|14	6: 8|48	6: 5|30	5: 2|10
2: 9|18	2: 5|10	3: 3|9	9: 8|72	5: 7|35	10: 2|20	4: 9|36
3: 7|21	8: 5|40	9: 2|18	5: 9|45	4: 2|8	2: 3|6	5: 5|25
6: 2|12	1: 9|9	1: 3|3	10: 5|50	8: 8|64		

6

9: 4|36	3: 10|30	7: 4|28	10: 4|40	5: 10|50	6: 4|24	2: 10|20
4: 10|40	6: 4|24	8: 4|32	3: 10|30	7: 4|28	5: 10|50	9: 4|36
7: 4|28	2: 10|20	10: 4|40	10: 10|100	8: 4|32	1: 10|10	0: 4|0

Lesson 38 (continued)

7

A 271 rounds to 27 tens.
B 350 rounds to 35 tens.
C 888 rounds to 89 tens.
D 695 rounds to 70 tens.
E 704 rounds to 70 tens.

8 Write the facts with no remainders.

A [shaded box] 7|31 — 4: 7|28
B [shaded box] 7|24 — 3: 7|21
C [shaded box] 7|5 — 0: 7|0
D [shaded box] 7|18 — 2: 7|14
E [shaded box] 7|10 — 1: 7|7
F [shaded box] 7|38 — 5: 7|35

9 Multiply to find the number you subtract. Then find the first remainder.

A 4: 24|1032, −96, 7
B 5: 53|2864, −265, 21
C 2: 27|5932, −54, 5
D 5: 42|2308, −210, 20
E 5: 37|189, −185, 4

10

A 240: 3|721, −6, 12, −12, 1
B 114: 5|570, −5, 7, −5, 20, −20, 0
C 876: 8|7009, −64, 60, −56, 49, −48, 1
D 105: 7|735, −7, 35, −35, 0
E 30: 2|61, −6, 1

Lesson 38 (continued)

11

A A machine has been making boxes for 7 hours. The machine makes 214 boxes each hour. How many boxes has the machine made?

214 × 7 = 1498

[1498] boxes

B Each time Ella fixes a suit, she uses 7 pins. Ella has used 84 pins. How many suits has Ella fixed?

7⟌84 = 12; −7, 14, −14, 0

[12] suits

C A factory makes 9 cars every hour. The factory made 1080 cars. How many hours did it take to make the cars?

9⟌1080 = 120; −9, 18, −18, 0

[120] hours

D Each student collected 7 cans. There are 1425 students in our school. How many cans were collected in all?

1425 × 7 = 9975

[9975] cans

Lesson 39

Test + Facts + Problems + Bonus = TOTAL

1

A	10⟌60 = [6]	10⟌60 = 6	6⟌60 = 10
B	10⟌40 = [4]	10⟌40 = 4	4⟌40 = 10
C	10⟌100 = [10]	10⟌100 = 10	
D	10⟌30 = [3]	10⟌30 = 3	3⟌30 = 10
E	10⟌70 = [7]	10⟌70 = 7	7⟌70 = 10
F	10⟌50 = [5]	10⟌50 = 5	5⟌50 = 10
G	10⟌90 = [9]	10⟌90 = 9	9⟌90 = 10
H	10⟌10 = [1]	10⟌10 = 1	1⟌10 = 10
I	10⟌80 = [8]	10⟌80 = 8	8⟌80 = 10
J	10⟌20 = [2]	10⟌20 = 2	2⟌20 = 10

2

A 62⟌256 = ~~3~~ 4; −186, 70 — 62⟌256 = 4; −248, 8

B 25⟌170 = 6; −150, 20 — [ ]

C 43⟌262 = ~~5~~ 6; −215, 47 — 43⟌262 = 6; −258, 4

D 38⟌152 = ~~3~~ 4; −114, 38 — 38⟌152 = 4; −152, 0

3

A 41⟌350 = 8; −328, 22 — [ ]

B 26⟌142 = 5; −130, 12 — [ ]

Part 3 continues on the next page.

Lesson 39 (continued)

C 37⟌218 = ~~6~~ 5; −222 — 37⟌218 = 5; −185, 33

D 18⟌70 = ~~4~~ 3; −72 — 18⟌70 = 3; −54, 16

4

A	4⟌32 = [8]	4⟌32 = 8	8⟌32 = 4
B	4⟌36 = [9]	4⟌36 = 9	9⟌36 = 4
C	4⟌24 = [6]	4⟌24 = 6	6⟌24 = 4
D	4⟌40 = [10]	4⟌40 = 10	10⟌40 = 4
E	4⟌28 = [7]	4⟌28 = 7	7⟌28 = 4

5

A [5⟌33 = 6] 54⟌326 = 6

B [3⟌7 = 2] 29⟌739 = 2

C [3⟌20 = 6] 33⟌2021 = 6

D [3⟌5 = 1] 32⟌505 = 1

E [2⟌3 = 1] 18⟌328 = 1

6

A 704 rounds to 70 tens.
B 919 rounds to 92 tens.
C 592 rounds to 59 tens.
D 495 rounds to 50 tens.
E 655 rounds to 66 tens.
F 837 rounds to 84 tens.

Lesson 39 (continued)

7

7⟌28 = 4	9⟌36 = 4	2⟌20 = 10	5⟌35 = 7	3⟌3 = 1	2⟌10 = 5	5⟌15 = 3
3⟌12 = 4	2⟌4 = 2	7⟌7 = 1	8⟌40 = 5	9⟌18 = 2	5⟌45 = 9	2⟌12 = 6
2⟌16 = 8	8⟌72 = 9	5⟌25 = 5	2⟌2 = 1	5⟌10 = 2	7⟌35 = 5	5⟌50 = 10
5⟌30 = 6	9⟌27 = 3	2⟌18 = 9	7⟌14 = 2	8⟌48 = 6	2⟌8 = 4	5⟌20 = 4
7⟌21 = 3	8⟌80 = 10	2⟌6 = 3	8⟌56 = 7	3⟌9 = 3	9⟌45 = 5	5⟌40 = 8
8⟌64 = 8	3⟌15 = 5	3⟌6 = 2	9⟌9 = 1	2⟌14 = 7		

8

10⟌40 = 4	4⟌28 = 7	10⟌70 = 7	4⟌36 = 9	10⟌90 = 9	4⟌24 = 6	10⟌20 = 2
10⟌100 = 10	4⟌24 = 6	10⟌10 = 1	4⟌40 = 10	4⟌32 = 8	10⟌70 = 7	10⟌100 = 10
4⟌28 = 7	10⟌80 = 8	10⟌60 = 6	4⟌28 = 7	10⟌30 = 3	10⟌80 = 8	4⟌32 = 8

9

Write the facts with no remainders.

A 7⟌13 [shaded] — 7⟌7 = 1
B 7⟌39 [shaded] — 7⟌35 = 5
C 7⟌9 [shaded] — 7⟌7 = 1
D 7⟌30 [shaded] — 7⟌28 = 4
E 7⟌23 [shaded] — 7⟌21 = 3
F 7⟌19 [shaded] — 7⟌14 = 2

Lesson 39 (continued)

10

A. 80 — 8⟌643, −64, 3

B. 687 — 2⟌1375, −12, 17, −16, 15, −14, 1

C. 103 — 9⟌927, −9, 27, −27, 0

D. 570 — 5⟌2853, −25, 35, −35, 03

E. 503 — 7⟌3524, −35, 24, −21, 3

11

Multiply to find the number you subtract. Then find the first remainder.

A. 6 — 24⟌1582, −144, 14

B. 5 — 16⟌8243, −80, 2

C. 2 — 46⟌983, −92, 6

D. 5 — 38⟌2039, −190, 13

12

A. Our family eats 3 cans of beans every week. We have 72 cans of beans. How many weeks will the beans last?
24 weeks
(24 — 3⟌72, −6, 12, −12, 0)

B. Alex made 3 rings every hour. He made 1320 rings. How many hours did Alex make rings?
440 hours
(440 — 3⟌1320, −12, 12, −12, 0)

C. The pilot practiced for 3 hours every day. She practiced for 75 hours. How many days did the pilot practice?
25 days
(25 — 3⟌75, −6, 15, −15, 0)

D. Lola used 5 batteries every month. She used batteries for 315 months. How many batteries did Lola use?
1575 batteries
(315 × 5 = 1575)

Lesson 40

Facts + Problems + Bonus = TOTAL

1

A. 6⟌24 B. 6⟌12 C. 6⟌18 D. 6⟌30

2

A. 1 — 6⟌6; 6⟌6 = 1; 1⟌6 = 6
B. 2 — 6⟌12; 6⟌12 = 2; 2⟌12 = 6
C. 3 — 6⟌18; 6⟌18 = 3; 3⟌18 = 6
D. 4 — 6⟌24; 6⟌24 = 4; 4⟌24 = 6
E. 5 — 6⟌30; 6⟌30 = 5; 5⟌30 = 6

3

A. 4⟌24 = 6 B. 4⟌36 = 9 C. 4⟌32 = 8 D. 4⟌28 = 7

4

A. 9 — 10⟌90; 10⟌90 = 9; 9⟌90 = 10
B. 1 — 10⟌10; 10⟌10 = 1; 1⟌10 = 10
C. 5 — 10⟌50; 10⟌50 = 5; 5⟌50 = 10
D. 2 — 10⟌20; 10⟌20 = 2; 2⟌20 = 10
E. 8 — 10⟌80; 10⟌80 = 8; 8⟌80 = 10
F. 3 — 10⟌30; 10⟌30 = 3; 3⟌30 = 10
G. 7 — 10⟌70; 10⟌70 = 7; 7⟌70 = 10
H. 4 — 10⟌40; 10⟌40 = 4; 4⟌40 = 10
I. 10 — 10⟌100; 10⟌100 = 10
J. 6 — 10⟌60; 10⟌60 = 6; 6⟌60 = 10

Lesson 40 (continued)

5

A. 6 (crossed out) — 29⟌178, −145, 33; 6 — 29⟌178, −174, 4

B. 4 — 35⟌168, −140, 28

C. 6 (crossed out) — 25⟌150, −125, 25; 6 — 25⟌150, −150, 0

6

A. 8 — 46⟌374, −368, 6

B. 6 — 24⟌167, −144, 23

C. 4 (crossed out) — 36⟌168, −180; 4 — 36⟌168, −144, 24

D. 8 (crossed out) — 31⟌274, −279; 8 — 31⟌274, −248, 26

7

A. 4 — 87⟌391; 9⟌39 = 4

B. 7 — 54⟌3828; 5⟌38 = 7

C. 2 — 26⟌847; 3⟌8 = 2

D. 8 — 75⟌6982; 8⟌70 = 8

8

4⟌32 = 8, 8⟌56 = 7, 7⟌28 = 4, 5⟌40 = 8, 2⟌18 = 9, 5⟌35 = 7, 9⟌27 = 3

2⟌12 = 6, 3⟌15 = 5, 8⟌48 = 6, 4⟌24 = 6, 9⟌36 = 4, 2⟌6 = 3, 5⟌25 = 5

Part 8 continues on the next page.

Lesson 40 (continued)

9⟌45 = 5, 2⟌16 = 8, 3⟌9 = 3, 2⟌20 = 10, 7⟌21 = 3, 8⟌64 = 8, 4⟌36 = 9

8⟌40 = 5, 4⟌28 = 7, 5⟌10 = 2, 2⟌4 = 2, 9⟌18 = 2, 3⟌12 = 4, 8⟌72 = 9

2⟌14 = 7, 3⟌6 = 2, 7⟌14 = 2, 5⟌15 = 3, 5⟌30 = 6, 4⟌40 = 10, 5⟌20 = 4

2⟌8 = 4, 7⟌35 = 5, 8⟌80 = 10, 2⟌10 = 5, 5⟌45 = 9

9

10⟌90 = 9, 6⟌12 = 2, 10⟌60 = 6, 10⟌20 = 2, 6⟌18 = 3, 10⟌70 = 7, 10⟌50 = 5

10⟌100 = 10, 6⟌18 = 3, 10⟌80 = 8, 6⟌12 = 2, 10⟌90 = 9, 10⟌60 = 6, 10⟌40 = 4

10⟌10 = 1, 10⟌100 = 10, 6⟌18 = 3, 10⟌70 = 7, 6⟌12 = 2, 10⟌30 = 3, 10⟌80 = 8

10

A. 233 rounds to 23 tens.
B. 116 rounds to 12 tens.
C. 384 rounds to 38 tens.
D. 808 rounds to 81 tens.
E. 195 rounds to 20 tens.
F. 497 rounds to 50 tens.

11

Write the facts with no remainders.

A. 7⟌17 — 7⟌14 = 2
B. 7⟌33 — 7⟌28 = 4
C. 7⟌26 — 7⟌21 = 3
D. 7⟌12 — 7⟌7 = 1
E. 7⟌3 — 7⟌0 = 0
F. 7⟌37 — 7⟌35 = 5

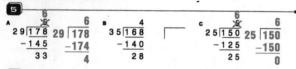

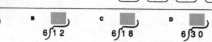

12

A. 260; 5)1302; −10; 30; −30; 2

B. 400; 2)801; −8; 01

C. 141; 3)423; −3; 12; −12; 3; −3; 0

D. 1504; 5)7524; −5; 25; −25; 24; −20; 4

13

Multiply to find the number you subtract. Then find the first remainder.

A. 2; 17)3894; −34; 4

B. 4; 52)248; −208; 40

C. 4; 24)9853; −96; 2

D. 1; 83)1452; −83; 62

14

A Each watch needed 4 wheels. There were 412 wheels. How many watches could be made?
103 watches
(103; 4)412; −4; 012; −12; 0)

B Mr. Singu worked 1232 hours. He made 4 toys each hour he worked. How many toys did Mr. Singu make?
4928 toys
(1232 × 4 = 4928)

C There were 7 men connecting wires. 308 wires were connected. How many wires did each man connect?
44 wires
(44; 7)308; −28; 28; −28; 0)

D Each pie needed 15 apples. Amy made 3 pies. How many apples did Amy use?
45 apples
(15 × 3 = 45)

Facts + Problems + Bonus = TOTAL

1

A. 6)18 B. 6)30 C. 6)12 D. 6)24

2

A. [1] 6)6; 1: 6)6; 6: 1)6

B. [2] 6)12; 2: 6)12; 6: 2)12

C. [3] 6)18; 3: 6)18; 6: 3)18

D. [4] 6)24; 4: 6)24; 6: 4)24

E. [5] 6)30; 5: 6)30; 6: 5)30

3

A. 2: 6)12 B. 3: 4)12

4

A. 7: 10)70 B. 2: 10)20 C. 5: 10)50 D. 1: 10)10

5

A. 37)132: ~~4~~ 3; −148 | 3: 37)132; −111; 21

B. 5: 41)210; −205; 5

C. 24)38: ~~2~~ 1; −48 | 1: 24)38; −24; 14

D. 29)170: ~~6~~ 5; −174 | 5: 29)170; −145; 25

6

A. 9: 83)759; −747; 12

B. 7: 62)499; −434; 65 | 8: 62)499; −496; 3

C. 8: 58)452; −464 | 7: 58)452; −406; 46

D. 8: 71)640; −568; 72 | 9: 71)640; −639; 1

7

A. [7: 4)31] 7: 41)3129

B. [6: 2)13] 6: 19)1331

C. [2: 9)25] 2: 85)245

D. [4: 7)28] 4: 67)2836

E. [1: 5)7] 1: 54)681

F. [5: 8)44] 5: 81)4444

G. [8: 4)35] 8: 38)3527

8

8: 4)32; 9: 2)18; 8: 10)80; 4: 7)28; 8: 2)16; 10: 10)100; 8: 8)64

2: 10)20; 7: 4)28; 6: 8)48; 5: 9)45; 4: 10)40; 3: 7)21; 9: 4)36

Part 8 continues on the next page.

9: 5)45; 1: 10)10; 5: 8)40; 6: 2)12; 6: 10)60; 2: 9)18; 8: 5)40

4: 9)36; 5: 10)50; 7: 5)35; 2: 7)14; 7: 2)14; 4: 2)8; 7: 10)70

4: 5)20; 3: 9)27; 3: 10)30; 6: 4)24; 5: 7)35; 9: 10)90; 9: 8)72

5: 5)25; 7: 8)56; 8: 4)32; 6: 5)30; 10: 4)40

9

3: 10)30; 3: 6)18; 7: 10)70; 2: 10)20; 2: 6)12; 3: 6)18; 10: 10)100

3: 6)18; 8: 10)80; 1: 10)10; 2: 6)12; 6: 10)60; 9: 10)90; 4: 10)40

10

Write the facts with no remainders.

A. 4)30 | 7: 4)28

B. 4)41 | 10: 4)40

C. 4)26 | 6: 4)24

D. 4)33 | 8: 4)32

E. 4)35 | 8: 4)32

F. 4)38 | 9: 4)36

11

205: 9)1845; −18; 45; −45; 0

980: 4)3921; −36; 32; −32; 1

352: 3)1056; −9; 15; −15; 6; −6; 0

670: 8)5367; −48; 56; −56; 7

1030: 7)7211; −7; 21; −21; 1

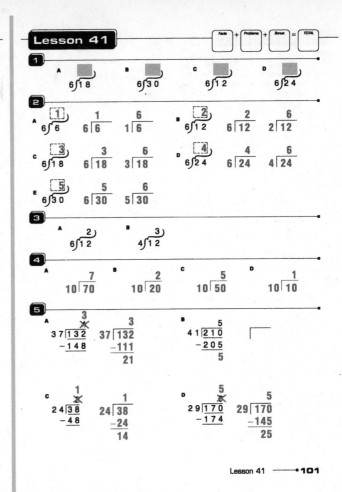

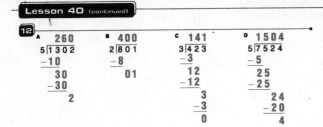

12

A Mrs. Smith bought 7 rolls of string. She can tie 84 packages with each roll. How many packages can she tie with the string she bought?

[588] packages

84
× 7
588

B 4 women are holding each rope. There are 36 women in all. How many ropes are there?

[9] ropes

4 | 36 = 9

C Jim does 2 sit-ups whenever he exercises. He has done 90 sit-ups. How many times has Jim exercised?

[45] times

2 | 90 = 45
−8
10
−10
0

13

Multiply to find the number you subtract. Then find the remainder.

A	B	C	D
24 | 9185 = 3	82 | 4238 = 5	25 | 108 = 4	37 | 1849 = 4
−72	−410	−100	−148
19	13	8	36

Test + Facts + Problems + Bonus = TOTAL

1

A [3] 6)18 6 | 18 = 3 3 | 18 = 6

B [4] 6)24 6 | 24 = 4 4 | 24 = 6

C [2] 6)12 6 | 12 = 2 2 | 12 = 6

D [5] 6)30 6 | 30 = 5 5 | 30 = 6

E [1] 6)6 6 | 6 = 1 1 | 6 = 6

2

A 9)18 = 2

B 3)18 = 6

3

A 65 | 505 = 7
−455
50

B 96 | 304 = 2
−192
112
96 | 304 = 3
−288
16

C 58 | 431 = 8
−464
58 | 431 = 7
−406
25

D 37 | 129 = 4
−148
37 | 129 = 3
−111
18

E 46 | 276 = 5
−230
46
46 | 276 = 6
−276
0

F 52 | 298 = 6
−312
52 | 298 = 5
−260
38

4

A 21 | 868 = 41
−84
28
−21
7

[2 | 9 = 4]
[2 | 3 = 1]

B 33 | 743 = 22
−66
83
−66
17

[3 | 7 = 2]
[3 | 8 = 2]

5

4 | 32 = 8	10 | 100 = 10	7 | 14 = 2	5 | 25 = 5	10 | 30 = 3	8 | 40 = 5	9 | 45 = 5
2 | 14 = 7	9 | 18 = 2	10 | 80 = 8	3 | 15 = 5	2 | 6 = 3	8 | 64 = 8	5 | 20 = 4
7 | 35 = 5	10 | 50 = 5	2 | 16 = 8	8 | 56 = 7	10 | 20 = 2	5 | 35 = 7	7 | 21 = 3
4 | 24 = 6	2 | 12 = 6	5 | 40 = 8	9 | 27 = 3	4 | 36 = 9	3 | 12 = 4	10 | 90 = 9
2 | 8 = 4	10 | 60 = 6	9 | 36 = 4	4 | 28 = 7	8 | 72 = 9	10 | 40 = 4	7 | 28 = 4
2 | 18 = 9	8 | 48 = 6	3 | 9 = 3	10 | 70 = 7	2 | 10 = 5		

6

6 | 24 = 4	4 | 32 = 8	6 | 18 = 3	6 | 30 = 5	4 | 24 = 6	6 | 12 = 2	6 | 30 = 5
4 | 28 = 7	6 | 12 = 2	10 | 30 = 3	10 | 70 = 7	6 | 24 = 4	4 | 36 = 9	6 | 18 = 3

7

A Fran brought 2 worms every time she went fishing. She went fishing 134 times. How many worms did Fran bring?

[268] worms

134
× 2
268

B Tina steered the ship 4 hours a day. She steered the ship for 120 hours. How many days did Tina steer the ship?

[30] days

4 | 120 = 30
−12
0

Part 7 continues on the next page.

C Every zoo I visited had 7 lions. I saw 140 lions. How many zoos did I visit?

[20] zoos

7 | 140 = 20
−14
0

D Every time Leroy went skiing, he went down 9 hills. He went skiing 216 times. How many hills did Leroy go down?

[1944] hills

216
× 9
1944

8

A 4 | 2830 = 707
−28
030
−28
2

B 7 | 861 = 123
−7
16
−14
21
−21
0

C 2 | 1994 = 997
−18
19
−18
14
−14
0

D 8 | 5672 = 709
−56
072
−72
0

E 5 | 1600 = 320
−15
10
−10
0

9

Write the facts with no remainders.

A 4 | 37 → 4 | 36 = 9	B 4 | 39 → 4 | 36 = 9	C 4 | 27 → 4 | 24 = 6
D 4 | 29 → 4 | 28 = 7	E 4 | 42 → 4 | 40 = 10	F 4 | 34 → 4 | 32 = 8
G 10 | 46 → 10 | 40 = 4	H 10 | 21 → 10 | 20 = 2	I 10 | 85 → 10 | 80 = 8
J 10 | 97 → 10 | 90 = 9	K 10 | 32 → 10 | 30 = 3	L 10 | 104 → 10 | 100 = 10

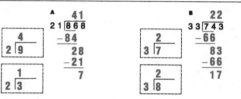

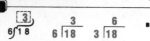

10 Multiply to find the number you subtract. Then find the remainder.

- A: 36 ⟌ 9245 — quotient 2; −72; remainder 20
- B: 58 ⟌ 1836 — quotient 3; −174; remainder 9
- C: 19 ⟌ 2134 — quotient 1; −19; remainder 2
- D: 42 ⟌ 2000 — quotient 4; −168; remainder 32

Lesson 43

Facts + Problems + Bonus = TOTAL

1

- A: 7 ⟌ 56
- B: 7 ⟌ 70
- C: 7 ⟌ 63
- D: 7 ⟌ 42

2

- A: 7 ⟌ 42 → 6; 7 ⟌ 42 = 6; 6 ⟌ 42 = 7
- B: 7 ⟌ 49 → 7; 7 ⟌ 49 = 7
- C: 7 ⟌ 56 → 8; 7 ⟌ 56 = 8; 8 ⟌ 56 = 7
- D: 7 ⟌ 63 → 9; 7 ⟌ 63 = 9; 9 ⟌ 63 = 7
- E: 7 ⟌ 70 → 10; 7 ⟌ 70 = 10; 10 ⟌ 70 = 7

3

- A: 38 ⟌ 186 — 5; −190 | 38 ⟌ 186 — 4; −152; 34
- B: 52 ⟌ 275 — 4; −208; 67 | 52 ⟌ 275 — 5; −260; 15
- C: 46 ⟌ 129 — 3; −138 | 46 ⟌ 129 — 2; −92; 37
- D: 26 ⟌ 161 — 7; −182 | 26 ⟌ 161 — 6; −156; 5
- E: 46 ⟌ 277 — 5; −230; 47 | 46 ⟌ 277 — 6; −276; 1

4

- A: 6 ⟌ 30 → 5; 6 ⟌ 30 = 5; 5 ⟌ 30 = 6
- B: 6 ⟌ 6 → 1; 6 ⟌ 6 = 1; 1 ⟌ 6 = 6
- C: 6 ⟌ 24 → 4; 6 ⟌ 24 = 4; 4 ⟌ 24 = 6
- D: 6 ⟌ 18 → 3; 6 ⟌ 18 = 3; 3 ⟌ 18 = 6
- E: 6 ⟌ 12 → 2; 6 ⟌ 12 = 2; 2 ⟌ 12 = 6

5

- 8 ⟌ 45 = 5; 8 ⟌ 64 = 8
- A: 77 ⟌ 4492 — 58; −385; 642; −616; 26
- 9 ⟌ 28 = 3; 9 ⟌ 25 = 2
- B: 86 ⟌ 2827 — 32; −258; 247; −172; 75

6

4 ⟌ 24 = 6	8 ⟌ 56 = 7	7 ⟌ 28 = 4	5 ⟌ 10 = 2	1 ⟌ 9 = 9	2 ⟌ 16 = 8	3 ⟌ 9 = 3
5 ⟌ 30 = 6	10 ⟌ 40 = 4	9 ⟌ 27 = 3	8 ⟌ 48 = 6	5 ⟌ 45 = 9	4 ⟌ 28 = 7	5 ⟌ 50 = 10
8 ⟌ 64 = 8	8 ⟌ 0 = 0	4 ⟌ 36 = 9	3 ⟌ 12 = 4	10 ⟌ 50 = 5	2 ⟌ 18 = 9	9 ⟌ 36 = 4
3 ⟌ 15 = 5	2 ⟌ 8 = 4	9 ⟌ 18 = 2	4 ⟌ 32 = 8	4 ⟌ 40 = 10	7 ⟌ 35 = 5	3 ⟌ 6 = 2
7 ⟌ 21 = 3	5 ⟌ 15 = 3	10 ⟌ 60 = 6	8 ⟌ 40 = 5	2 ⟌ 14 = 7	9 ⟌ 45 = 5	1 ⟌ 3 = 3
4 ⟌ 0 = 0	2 ⟌ 10 = 5	10 ⟌ 10 = 1	8 ⟌ 72 = 9	2 ⟌ 20 = 10		

7

6 ⟌ 30 = 5	7 ⟌ 42 = 6	6 ⟌ 24 = 4	7 ⟌ 49 = 7	6 ⟌ 18 = 3	7 ⟌ 42 = 6	6 ⟌ 30 = 5
6 ⟌ 24 = 4	6 ⟌ 18 = 3	7 ⟌ 49 = 7	7 ⟌ 42 = 6	6 ⟌ 12 = 2	6 ⟌ 30 = 5	6 ⟌ 24 = 4

8 Write the facts with no remainders.

- A: 4 ⟌ 43; 4 ⟌ 40 = 10
- B: 4 ⟌ 31; 4 ⟌ 28 = 7
- C: 4 ⟌ 25; 4 ⟌ 24 = 6
- D: 4 ⟌ 38; 4 ⟌ 36 = 9
- E: 4 ⟌ 34; 4 ⟌ 32 = 8
- F: 4 ⟌ 30; 4 ⟌ 28 = 7

9

A It rained 5 centimeters of water every hour. It rained for 80 hours. How many centimeters of water did it rain?

400 centimeters (80 × 5 = 400)

B 4 people can sit at each table. There are 420 people sitting at tables. How many tables are there?

105 tables (4 ⟌ 420 = 105; −4; 020; −20; 0)

C I have 7 dishes on each shelf. If I have 140 dishes, how many shelves do I have?

20 shelves (7 ⟌ 140 = 20; −14; 0)

10

- A: 2 ⟌ 9601 — 4800; −8; 16; −16; 01
- B: 4 ⟌ 2435 — 608; −24; 35; −32; 3
- C: 2 ⟌ 2081 — 1040; −2; 08; −8; 1
- D: 7 ⟌ 2943 — 420; −28; 14; −14; 3
- E: 5 ⟌ 1984 — 396; −15; 48; −45; 34; −30; 4

11 Multiply to find the number you subtract. Then find the remainder.

- A: 15 ⟌ 1349 — 8; −120; 14
- B: 20 ⟌ 8372 — 4; −80; 3
- C: 47 ⟌ 985 — 2; −94; 4
- D: 42 ⟌ 1354 — 3; −126; 9

Lesson 44

Facts + Problems + Bonus = TOTAL

1

A 7)49 B 7)63 C 7)56 D 7)42

2

A 7)42 → [6]; 7)42 → 6; 6)42 → 7
B 7)49 → [7]; 7)49 → 7
C 7)56 → [8]; 7)56 → 8; 8)56 → 7
D 7)63 → [9]; 7)63 → 9; 9)63 → 7
E 7)70 → [10]; 7)70 → 10; 10)70 → 7

3

A 6)12 → 2 B 6)24 → 4 C 6)30 → 5 D 6)18 → 3

4

A 76)393 → 4; −304; 89 | 76)393 → 5; −380; 13
B 28)153 → 6; −168 | 28)153 → 5; −140; 13
C 47)130 → 3; −141 | 47)130 → 2; −94; 36
D 65)325 → 4; −260; 65 | 65)325 → 5; −325; 0
E 33)247 → 8; −264 | 33)247 → 7; −231; 16
F 46)277 → 5; −230; 47 | 46)277 → 6; −276; 1

Lesson 44 (continued)

5

A [6)19 → 3] [6)13 → 2] 58)1872 → 32; −174; 132; −116; 16
B [7)10 → 1] [7)26 → 3] 72)984 → 13; −72; 264; −216; 48
C [5)9 → 1] [5)47 → 9] 48)948 → 19; −48; 468; −432; 36
D [6)26 → 4] [6)8 → 1] 62)2561 → 41; −248; 81; −62; 19

6

6)30 → 5 | 9)18 → 2 | 7)28 → 4 | 3)15 → 5 | 4)28 → 7 | 8)48 → 6 | 7)7 → 1
5)5 → 1 | 2)12 → 6 | 4)24 → 6 | 8)56 → 7 | 6)12 → 2 | 3)9 → 3 | 9)45 → 5
5)35 → 7 | 6)6 → 1 | 2)4 → 2 | 7)21 → 3 | 9)36 → 4 | 5)25 → 5 | 4)40 → 10
3)12 → 4 | 4)32 → 8 | 9)27 → 3 | 6)18 → 3 | 2)14 → 7 | 8)80 → 10 | 7)35 → 5
10)20 → 2 | 8)72 → 9 | 7)14 → 2 | 8)40 → 5 | 5)40 → 8 | 2)6 → 3 | 6)24 → 4
5)20 → 4 | 10)100 → 10 | 8)64 → 8 | 3)6 → 2 | 4)36 → 9

Lesson 44 (continued)

7

7)49 → 7 | 6)18 → 3 | 7)42 → 6 | 4)28 → 7 | 6)6 → 1 | 7)49 → 7 | 4)32 → 8
7)42 → 6 | 6)12 → 2 | 4)24 → 6 | 7)49 → 7 | 6)30 → 5 | 7)42 → 6 | 4)36 → 9

8

A Mr. and Mrs. Kato danced 8 dances every night. They danced 280 dances. How many nights did Mr. and Mrs. Kato go dancing?

[35] nights

8)280 → 35; −24; 40; −40; 0

B A squirrel cracked 8 nuts every minute. It cracked nuts for 40 minutes. How many nuts did the squirrel crack?

[320] nuts

40 × 8 = 320

C Beth took 6 pictures every hour. She took 192 pictures. How many hours did Beth take pictures?

[32] hours

6)192 → 32; −18; 12; −12; 0

9

A 6)8523 → 1420; −6; 25; −24; 12; −12; 3
B 7)1490 → 212; −14; 9; −7; 20; −14; 6
C 5)2300 → 460; −20; 30; −30; 0
D 4)2436 → 609; −24; 36; −36; 0

Lesson 45

Facts + Problems + Bonus = TOTAL

1

A 4)8 B 4)20 C 4)12 D 4)16

2

A 4)4 → [1]; 4)4 → 1; 1)4 → 4
B 4)8 → [2]; 4)8 → 2; 2)8 → 4
C 4)12 → [3]; 4)12 → 3; 3)12 → 4
D 4)16 → [4]; 4)16 → 4
E 4)20 → [5]; 4)20 → 5; 5)20 → 4

3

A 2)20 → 10 B 4)20 → 5

4

A 7)56 → [8]; 7)56 → 8; 8)56 → 7
B 7)63 → [9]; 7)63 → 9; 9)63 → 7
C 7)49 → [7]; 7)49 → 7
D 7)70 → [10]; 7)70 → 10; 10)70 → 7
E 7)42 → [6]; 7)42 → 6; 6)42 → 7

5

A 73)215 → 3; −219 | 73)215 → 2; −146; 69
B 54)330 → 5; −270; 60 | 54)330 → 6; −324; 6

Part 5 continues on the next page.

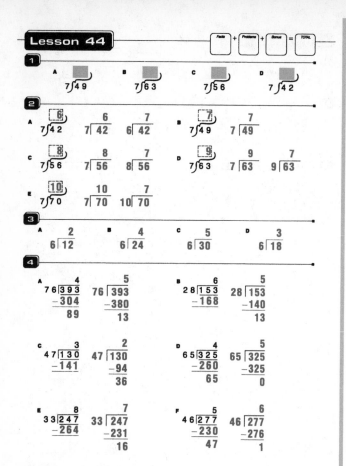

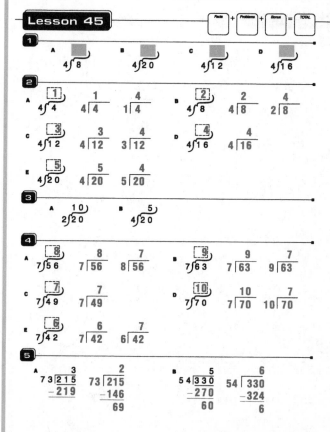

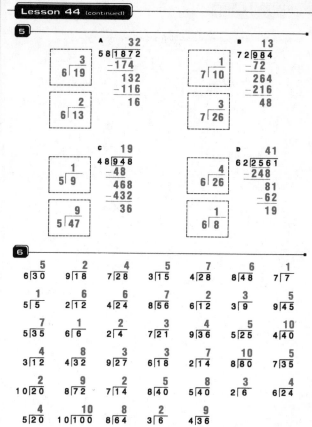

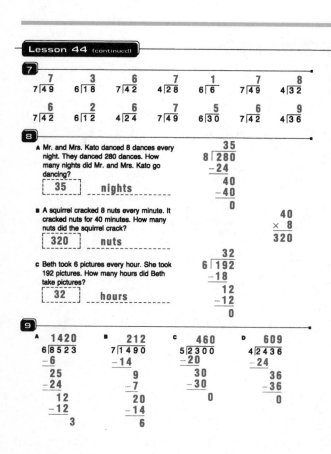

Lesson 45 (continued)

C 46)419 = 8; −368; 51 46)419 = 9; −414; 5

D 37)255 = 7; −259 37)255 = 6; −222; 33

E 37)94 = 3; −111 37)94 = 2; −74; 20

F 43)217 = 4; −172; 45 43)217 = 5; −215; 2

6

A 6)8 = 1; 6)18 = 3 | 57)752 = 13; −57; 182; −171; 11

B 2)9 = 4; 2)7 = 3 | 22)946 = 43; −88; 66; −66; 0

C 4)25 = 6; 4)19 = 4 | 39)2528 = 64; −234; 188; −156; 32

D 7)17 = 2; 7)33 = 4 | 70)1726 = 24; −140; 326; −280; 46

7

7)35 = 5, 8)56 = 7, 3)3 = 1, 6)12 = 2, 4)36 = 9, 3)9 = 3, 8)72 = 9

9)45 = 5, 4)24 = 6, 5)15 = 3, 8)48 = 6, 2)8 = 4, 6)30 = 5, 7)14 = 2

Part 7 continues on the next page.

Lesson 45 (continued)

2)18 = 9, 9)36 = 4, 6)6 = 1, 7)21 = 3, 8)40 = 5, 4)28 = 7, 2)10 = 5

8)64 = 8, 7)28 = 4, 9)27 = 3, 3)6 = 2, 6)24 = 4, 3)12 = 4, 4)40 = 10

9)18 = 2, 6)18 = 3, 2)2 = 1, 4)32 = 8, 5)30 = 6, 2)16 = 8, 5)50 = 10

5)35 = 7, 3)15 = 5, 5)20 = 4, 7)7 = 1, 5)25 = 5

8

7)56 = 8, 7)42 = 6, 4)8 = 2, 7)70 = 10, 7)56 = 8, 4)12 = 3, 7)49 = 7

7)63 = 9, 7)70 = 10, 7)42 = 6, 4)8 = 2, 7)63 = 9, 7)56 = 8, 7)70 = 10

7)49 = 7, 4)12 = 3, 7)63 = 9, 4)12 = 3, 4)8 = 2

9

A 6)1949 = 324; −18; 14; −12; 29; −24; 5

B 8)5675 = 709; −56; 75; −72; 3

C 2)1119 = 559; −10; 11; −10; 19; −18; 1

D 6)9084 = 1514; −6; 30; −30; 8; −6; 24; −24; 0

E 7)2135 = 305; −21; 35; −35; 0

10 Write the facts with no remainders.

A 6)9 ▢ 6)6 = 1 **B** 6)32 ▢ 6)30 = 5 **C** 6)20 ▢ 6)18 = 3

D 6)3 ▢ 6)0 = 0 **E** 6)17 ▢ 6)12 = 2 **F** 6)26 ▢ 6)24 = 4

Lesson 46

Facts + Problems + Bonus = TOTAL

1

A 4)12 ▢ **B** 4)16 ▢ **C** 4)8 ▢ **D** 4)20 ▢

2

A 4)4 [1]; 4)4 = 1; 1)4 = 4 **B** 4)8 [2]; 4)8 = 2; 2)8 = 4

C 4)12 [3]; 4)12 = 3; 3)12 = 4 **D** 4)16 [4]; 4)16 = 4

E 4)20 [5]; 4)20 = 5; 5)20 = 4

3

A 3)12 = 4 **B** 2)12 = 6

4

A 7)63 [9]; 7)63 = 9; 9)63 = 7 **B** 7)49 [7]; 7)49 = 7

C 7)42 [6]; 7)42 = 6; 6)42 = 7 **D** 7)70 [10]; 7)70 = 10; 10)70 = 7

E 7)56 [8]; 7)56 = 8; 8)56 = 7

5

A 2)8; 2)13 | 22)794 = 36; −66; 134; −132; 2

B 4)32; 4)9 | 35)3243 = 92; −315; 93; −70; 23

C 6)23; 6)9 | 56)2327 = 41; −224; 87; −56; 31

Part 5 continues on the next page.

Lesson 46 (continued)

D 5)25; 5)44 | 51)2481 = 48; −204; 441; −408; 33

E 7)38; 7)23 | 72)3829 = 53; −360; 229; −216; 13

6

6)18 = 3, 9)36 = 4, 2)4 = 2, 8)40 = 5, 5)10 = 2, 4)24 = 6, 5)25 = 5

10)90 = 9, 5)40 = 8, 6)6 = 1, 3)9 = 3, 9)18 = 2, 4)40 = 10, 10)30 = 3

9)9 = 1, 4)32 = 8, 10)10 = 1, 6)30 = 5, 8)64 = 8, 5)15 = 3, 2)12 = 6

9)27 = 3, 7)14 = 2, 3)6 = 2, 8)72 = 9, 6)12 = 2, 7)28 = 4, 3)12 = 4

6)24 = 4, 2)12 = 6, 8)48 = 6, 2)8 = 4, 9)45 = 5, 4)28 = 7, 5)40 = 8

3)15 = 5, 8)56 = 7, 7)35 = 5, 4)36 = 9, 7)21 = 3

7

7)42 = 6, 4)12 = 3, 7)56 = 8, 7)63 = 9, 4)8 = 2, 7)49 = 7, 4)12 = 3

7)63 = 9, 7)56 = 8, 4)8 = 2, 7)42 = 6, 4)12 = 3, 7)70 = 10, 7)56 = 8

7)70 = 10, 4)8 = 2, 7)49 = 7, 7)63 = 9, 7)70 = 10

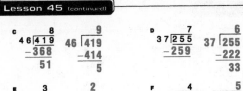

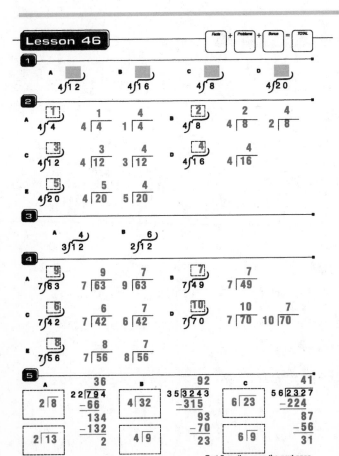

8 Change the answer if the remainder is too big or too small.

A. 27)114 → 3, −81, 33; corrected: 27)114 → 4, −108, 6

B. 56)473 → 8, −448, 25

C. 74)370 → 4, −296, 74; corrected: 74)370 → 5, −370, 0

D. 63)307 → 5, −315; corrected: 63)307 → 4, −252, 55

E. 48)392 → 7, −336, 56; corrected: 48)392 → 8, −384, 8

9 Write the facts with no remainders.

A. 6)22 → 6)18 = 3

B. 6)15 → 6)12 = 2

C. 6)27 → 6)24 = 4

D. 6)10 → 6)6 = 1

E. 6)33 → 6)30 = 5

F. 6)25 → 6)24 = 4

10

A. 4)3584 = 896: −32, 38, −36, 24, −24, 0

B. 9)933 = 103: −9, 33, −27, 6

C. 6)1381 = 230: −12, 18, −18, 01

D. 3)1230 = 410: −12, 3, −3, 00

E. 5)41,879 = 8375: −40, 18, −15, 37, −35, 29, −25, 4

11

A. Each box will hold 8 statues. There are 120 boxes. How many statues can be put in the boxes?
960 statues
120 × 8 = 960

B. There are 5 offices on each floor. There are 120 floors. How many offices are there?
600 offices
120 × 5 = 600

C. Art's CD company wants to make 2000 copies of a CD. His machine makes 5 CDs every minute. How many minutes will it take to make the CDs?
400 minutes
5)2000 = 400: −20, 0

D. Rose's company makes 525 sewing machines every day. They made sewing machines for 5 days. How many sewing machines did they make?
2625 sewing machines
525 × 5 = 2625

Lesson 47

Test + Facts + Problems + Bonus = TOTAL

1

A. 6)54 B. 6)36 C. 6)48 D. 6)42

2

A. 6)36 → 6; 6)36 = 6

B. 6)42 → 7; 6)42 = 7; 7)42 = 6

C. 6)48 → 8; 6)48 = 8; 8)48 = 6

D. 6)54 → 9; 6)54 = 9; 9)54 = 6

E. 6)60 → 10; 6)60 = 10; 10)60 = 6

3

A. 7)49 = 7 B. 7)42 = 6 C. 7)63 = 9 D. 7)56 = 8

4

A. 4)20 → 5; 4)20 = 5; 5)20 = 4

B. 4)4 → 1; 4)4 = 1; 1)4 = 4

C. 4)16 → 4; 4)16 = 4

D. 4)8 → 2; 4)8 = 2; 2)8 = 4

E. 4)12 → 3; 4)12 = 3; 3)12 = 4

5

A. 5)40, 5)30; 53)4014 = 75: −371, 304, −265, 39

B. 7)53, 7)13; 65)5328 = 81: −520, 128, −65, 63

Part 5 continues on the next page.

C. 2)6, 2)13; 24)607 = 25: −48, 127, −120, 7

D. 6)20, 6)12; 62)1984 = 32: −186, 124, −124, 0

6

A. 21)1542 = 73: −147, 72, −63, 9

B. 27)1234 = 45: −108, 154, −135, 19

C. 43)1824 = 42: −172, 104, −86, 18

D. 37)1268 = 34: −111, 158, −148, 10

7

4)24 = 6, 9)27 = 3, 6)18 = 3, 4)36 = 9, 7)70 = 10, 6)12 = 2, 5)10 = 2

10)50 = 5, 7)49 = 7, 4)40 = 10, 8)72 = 9, 6)30 = 5, 9)36 = 4, 3)6 = 2

7)35 = 5, 8)40 = 5, 5)30 = 6, 7)14 = 2, 9)45 = 5, 4)32 = 8, 7)56 = 8

7)21 = 3, 2)6 = 3, 9)18 = 2, 3)15 = 5, 10)40 = 4, 5)20 = 4, 7)42 = 6

8)56 = 7, 3)12 = 4, 7)63 = 9, 2)18 = 9, 6)24 = 4, 4)28 = 7, 8)48 = 6

3)9 = 3, 2)14 = 7, 5)35 = 7, 7)28 = 4, 8)64 = 8

Lesson 47 (continued)

8

4)16 = 4 | 6)36 = 6 | 6)42 = 7 | 4)16 = 4 | 4)20 = 5 | 4)12 = 3 | 4)16 = 4

6)36 = 6 | 4)12 = 3 | 4)8 = 2 | 6)42 = 7 | 4)20 = 5 | 4)8 = 2 | 6)36 = 6

9

Change the answer if the remainder is too big or too small.

A 66)194 = 3; −198 → 66)194 = 2; −132; 62

B 47)219 = 5; −235 → 47)219 = 4; −188; 31

C 53)134 = 2; −106; 28 → (blank box)

D 84)675 = 7; −588; 87 → 84)675 = 8; −672; 3

10

A A race horse ran 312 kilometers each week. It ran for 3 weeks. How many kilometers did it run?
312 × 3 = 936
936 kilometers

B 424 people want to take a train ride. Each seat on the train holds 4 people. How many seats are needed?
4)424 = 106; −4; 24; −24; 0
106 seats

C Angel, the car dealer, had 1920 cars. He put the same number of cars into each of 6 lots. How many cars did he put in each lot?
6)1920 = 320; −18; 12; −12; 00
320 cars

D It took Yoko 7 months to build a dam. 231 loads of rock were used each month. How many loads of rock went into the dam?
231 × 7 = 1617
1617 loads

Lesson 48

Facts + Problems + Bonus = TOTAL

1

A 6)48 B 6)60 C 6)42 D 6)54

2

A 6)36 = 6 | 6)36 = 6

B 6)42 = 7 | 6)42 = 7 | 7)42 = 6

C 6)48 = 8 | 6)48 = 8 | 8)48 = 6

D 6)54 = 9 | 6)54 = 9 | 9)54 = 6

E 6)60 = 10 | 6)60 = 10 | 10)60 = 6

3

A 8 people got on an empty bus. At the first bus stop, 1 person got off. How many people end up on the bus?
8 − 1 = 7
7 people

B The store has 27 cakes. A truck brings 35 cakes. How many cakes does the store end up with?
27 + 35 = 62
62 cakes

C Kip has 6 books. He gets 12. How many books does Kip end up with?
6 + 12 = 18
18 books

D Sid has 7 cans of paint. He uses 6. How many cans does Sid end up with?
7 − 6 = 1
1 cans

E Rob has 9 pens. He got 4. How many pens does Rob have now?
9 + 4 = 13
13 pens

F Leon had 40 nails. He used 5 nails. How many nails does Leon have now?
40 − 5 = 35
35 nails

Part 3 continues on the next page.

Lesson 48 (continued)

G A farmer had 135 beets. He eats 14. How many beets does the farmer end up with?
135 − 14 = 121
121 beets

4

A 4)8 = 2 | 4)8 = 2 | 2)8 = 4

B 4)12 = 3 | 4)12 = 3 | 3)12 = 4

C 4)20 = 5 | 4)20 = 5 | 5)20 = 4

D 4)4 = 1 | 4)4 = 1 | 1)4 = 4

E 4)16 = 4 | 4)16 = 4

5

4)28 = 7 | 7)21 = 3 | 6)18 = 3 | 9)18 = 2 | 7)49 = 7 | 8)48 = 6 | 9)45 = 5

5)10 = 2 | 8)40 = 5 | 4)24 = 6 | 2)6 = 3 | 6)24 = 4 | 7)14 = 2 | 8)56 = 7

3)12 = 4 | 7)7 = 1 | 8)72 = 9 | 4)32 = 8 | 7)63 = 9 | 6)12 = 2 | 2)8 = 4

7)42 = 6 | 3)9 = 3 | 6)6 = 1 | 8)64 = 8 | 4)36 = 9 | 5)25 = 5 | 7)35 = 5

5)15 = 3 | 2)4 = 2 | 3)6 = 2 | 7)56 = 8 | 5)40 = 8 | 4)40 = 10 | 6)30 = 5

7)28 = 4 | 2)10 = 5 | 9)27 = 3 | 3)15 = 5 | 9)36 = 4

6

6)36 = 6 | 4)16 = 4 | 6)42 = 7 | 4)8 = 2 | 4)20 = 5 | 4)16 = 4 | 6)42 = 7

4)20 = 5 | 6)36 = 6 | 4)12 = 3 | 4)20 = 5 | 6)42 = 7 | 4)8 = 2 | 4)16 = 4

Lesson 48 (continued)

7

Write the facts with no remainders.

A 7)71 → 7)70 = 10

B 7)46 → 7)42 = 6

C 7)65 → 7)63 = 9

D 7)54 → 7)49 = 7

E 7)60 → 7)56 = 8

F 7)55 → 7)49 = 7

8

A 5)45, 5)22 — 54)4542 = 84; −432; 222; −216; 6

B 8)52, 8)55 — 78)5234 = 67; −468; 554; −546; 8

C 4)27, 4)20 — 42)2716 = 64; −252; 196; −168; 28

9

A 94)3065 = 32; −282; 245; −188; 57

B 7)2942 = 420; −28; 14; −14; 2

C 39)2972 = 76; −273; 242; −234; 8

D 8)8168 = 1021; −8; 16; −16; 8; −8; 0

10

Change the answer if the remainder is too big or too small.

A 26)120 = 5; −130 → 26)120 = 4; −104; 16

B 37)159 = 4; −148; 11 → (blank box)

C 14)38 = 3; −42 → 14)38 = 2; −28; 10

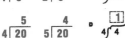

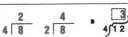

Lesson 49

Test + Facts + Problems + Bonus = TOTAL

1

- **A** [7] 6⟌42; 6⟌42 = 7; 7⟌42 = 6
- **B** [6] 6⟌36; 6⟌36 = 6
- **C** [9] 6⟌54; 6⟌54 = 9; 9⟌54 = 6
- **D** [10] 6⟌60; 6⟌60 = 10; 10⟌60 = 6
- **E** [8] 6⟌48; 6⟌48 = 8; 8⟌48 = 6

2

- **A** 4⟌20 = 5
- **B** 4⟌8 = 2
- **C** 4⟌16 = 4
- **D** 4⟌12 = 3

3

A A man has 25 tools. 2 are picks and the rest are saws. How many saws does he have?

[23] saws
25 − 2 = 23

B Greg has trees. 2 are pine trees and 38 are oak trees. How many trees does Greg have?

[40] trees
2 + 38 = 40

C Clara has tools. She has 34 drills and 9 hammers. How many tools does she have?

[43] tools
34 + 9 = 43

D Ryan has 9 flowers. He has 5 roses and the rest are tulips. How many tulips does Ryan have?

[4] tulips
9 − 5 = 4

E Andy has 219 buttons. He has 106 blue buttons and the rest are black. How many black buttons does Andy have?

[113] buttons
219 − 106 = 113

Part 3 continues on the next page.

Lesson 49 (continued)

F The farmer has animals. She has 40 pigs and 30 chickens. How many animals does the farmer have?

[70] animals
40 + 30 = 70

G Ms. Dodge's store has 54 chairs. There are 19 kitchen chairs and the rest are living room chairs. How many living room chairs does Ms. Dodge's store have?

[35] chairs
54 − 19 = 35

4

8⟌64 = 8	7⟌42 = 6	5⟌35 = 7	3⟌12 = 4	4⟌24 = 6	2⟌18 = 9	8⟌56 = 7
7⟌28 = 4	6⟌24 = 4	2⟌14 = 7	4⟌12 = 3	2⟌16 = 8	7⟌21 = 3	4⟌16 = 4
10⟌100 = 10	8⟌72 = 9	7⟌49 = 7	9⟌36 = 4	7⟌14 = 2	6⟌18 = 3	8⟌48 = 6
4⟌32 = 8	9⟌27 = 3	7⟌63 = 9	8⟌40 = 5	4⟌20 = 5	7⟌35 = 5	10⟌80 = 8
5⟌45 = 9	4⟌8 = 2	3⟌15 = 5	6⟌12 = 2	10⟌20 = 2	2⟌12 = 6	4⟌28 = 7
3⟌9 = 3	6⟌30 = 5	7⟌56 = 8	4⟌36 = 9	5⟌20 = 4		

5

6⟌42 = 7	6⟌60 = 10	6⟌54 = 9	6⟌36 = 6	4⟌16 = 4	6⟌48 = 8	6⟌54 = 9
4⟌20 = 5	6⟌60 = 10	6⟌36 = 6	6⟌54 = 9	6⟌48 = 8	6⟌42 = 7	6⟌60 = 10

Lesson 49 (continued)

6

Write the facts with no remainders.

- **A** 7⟌47; 7⟌42 = 6
- **B** 7⟌50; 7⟌49 = 7
- **C** 7⟌61; 7⟌56 = 8
- **D** 7⟌57; 7⟌56 = 8
- **E** 7⟌45; 7⟌42 = 6
- **F** 7⟌73; 7⟌70 = 10

7

A 7⟌914 = 130
```
 -7
 21
-21
  4
```

B 59⟌2494 = 42
```
-236
 134
-118
  16
```

C 3⟌7215 = 2405
```
 -6
 12
-12
 15
-15
  0
```

D 87⟌2847 = 32
```
-261
 237
-174
  63
```

8

A [9⟌22] [9⟌36] 94⟌2235 = 23
```
 188
 355
-282
  73
```

B [4⟌25] [ ] 36⟌252 = 7
```
-252
   0
```

C [5⟌14] [5⟌7] 46⟌1448 = 31
```
-138
  68
 -46
  22
```

D [2⟌8] [2⟌19] 21⟌817 = 38
```
 -63
 187
-168
  19
```


Lesson 50

Test + Facts + Problems + Bonus = TOTAL

1

- **A** [9] 6⟌54; 6⟌54 = 9; 9⟌54 = 6
- **B** [10] 6⟌60; 6⟌60 = 10; 10⟌60 = 6
- **C** [6] 6⟌36; 6⟌36 = 6
- **D** [8] 6⟌48; 6⟌48 = 8; 8⟌48 = 6
- **E** [7] 6⟌42; 6⟌42 = 7; 7⟌42 = 6

2

A [7⟌8] [7⟌53] 74⟌7934 = 107
```
 -74
 534
-518
  16
```

B [6⟌12] [6⟌48] 58⟌12,075 = 208
```
-116
 475
-464
  11
```

C [2⟌9] [2⟌10] 21⟌8501 = 404
```
-84
101
-84
 17
```

D [6⟌38] [6⟌59] 63⟌38,387 = 609
```
-378
 587
-567
  20
```

3

A A kid has 37 toothpicks. His friend has 110 toothpicks. How many more toothpicks does his friend have than he does?

[73] toothpicks
110 − 37 = 73

B Willie's turkey weighs 30 pounds. Alma's turkey weighs 25 pounds. How many pounds lighter is Alma's turkey than Willie's?

[5] pounds
30 − 25 = 5

Part 3 continues on the next page.

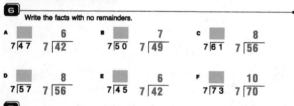

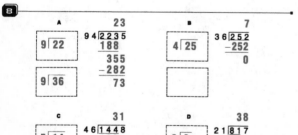

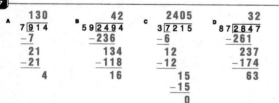

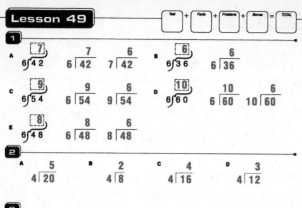

c Hole A is 22 feet deep. Hole B is 4 feet deeper than hole A. How many feet deep is hole B?

22
+4
26

26 feet

d Herb has 139 seeds in a bag. His mom has 657 more seeds than he does. How many seeds does his mom have?

139
+657
796

796 seeds

4

A The first story problem that Soo works deals with the same number again and again. + − × ÷

B The next problem that Soo works does not deal with the same number again and again. + − × ÷

C The next problem that Soo works does not deal with the same number again and again. + − × ÷

D The next problem that Soo works deals with the same number again and again. + − × ÷

E The next problem that Soo works does not deal with the same number again and again. + − × ÷

5

A 6)43 6)23

74
58)4293
−406
233
−232
1

B 2)8 2)7

32
24)789
−72
69
−48
21

C 7)50 7)34

75
67)5034
−469
344
−335
9

D 4)15 4)12

43
35)1524
−140
124
−105
19

6

2	7	3	4	3	9	5
7)14	8)56	4)12	9)36	9)27	4)36	5)25
3	9	8	9	2	5	5
6)18	7)63	5)40	8)72	3)6	4)20	2)10
10	2	3	2	5	2	2
5)50	4)8	7)21	2)4	7)35	5)10	9)18
7	8	2	5	5	10	6
4)28	7)56	6)12	8)40	6)30	2)20	4)24
3	7	8	8	8	4	4
3)9	7)49	2)16	4)32	8)64	7)28	3)12
6	4	6	5	4		
7)42	4)16	8)48	9)45	6)24		

7

8	10	9	4	7	8	10
6)48	6)60	6)54	4)16	6)42	6)48	6)60
9	2	6	10	3	9	7
6)54	4)8	6)36	6)60	4)12	6)54	6)42

8

Write the facts with no remainders.

A 7)58 8 7)56

B 7)52 7 7)49

C 7)65 9 7)63

D 7)43 6 7)42

E 7)74 10 7)70

F 7)68 9 7)63

G 4)10 2 4)8

H 4)7 1 4)4

I 4)18 4 4)16

Part 8 continues on the next page.

J 4)23 5 4)20

K 4)14 3 4)12

L 4)2 0 4)0

9

A
2304
4)9216
−8
12
−12
16
−16
0

B
48
74)3588
−296
628
−592
36

C
520
7)3642
−35
14
−14
02

D
13
71)946
−71
236
−213
23

E
896
4)3584
−32
38
−36
24
−24
0

F
103
9)933
−9
33
−27
6

G
230
6)1381
−12
18
−18
01

H
410
3)1230
−12
3
−3
00

Facts + Problems + Bonus = TOTAL

1

A 9)63 **B** 9)81 **C** 9)72 **D** 9)90

2

A 6 9)54 6 9)54 9 6)54

B 7 9)63 7 9)63 9 7)63

C 8 9)72 8 9)72 9 8)72

D 9 9)81 9 9)81

E 10 9)90 10 9)90 9 10)90

3

A 7 6)42 **B** 6 6)36 **C** 9 6)54 **D** 8 6)48

4

A Donna's store has 37 pairs of shoes. Mable's store has 110. How many more pairs of shoes does Mable's store have?

110
−37
73

73 pairs

B There were 25 ships in the harbor on Monday. There were 30 ships in the harbor on Tuesday. How many more ships were in the harbor on Tuesday than on Monday?

30
−25
5

5 ships

C There are 22 cars at stoplight A. There are 4 more cars at stoplight B than at A. How many cars are at stoplight B?

22
+4
26

26 cars

D Curt's oil well is 139 meters deep. He drills 657 more meters. How many meters deep is Curt's oil well?

139
+657
796

796 meters

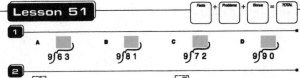

Lesson 51 (continued)

5

A

Boxes: 7|15, 7|33

74|15,125 = 204
−148
325
−296
29

B

Boxes: 6|38, 6|58

62|37,781 = 609
−372
581
−558
23

C

Boxes: 7|56, 7|48

69|55,684 = 807
−552
484
−483
1

D

Boxes: 2|7, 2|11

22|6714 = 305
−66
114
−110
4

6

A The first story problem that Sal works does not deal with the same number again and again. + − (underlined) / × ÷

B The next problem that Sal works deals with the same number again and again. + − / × ÷ (underlined)

C The next problem that Sal works deals with the same number again and again. + − / × ÷ (underlined)

7

A The problem does not deal with the same number again and again. The problem does not give the big number. (+ circled) − / × ÷

B The problem deals with the same number again and again. The problem does not give the big number. + − / (× circled) ÷

C The problem deals with the same number again and again. The problem gives the big number. + − / × (÷ circled)

D The problem does not deal with the same number again and again. The problem gives the big number. + (− circled) / × ÷

Part 7 continues on the next page.

Lesson 51 (continued)

E The problem deals with the same number again and again. The problem does not give the big number. + − / (× circled) ÷

F The problem deals with the same number again and again. The problem gives the big number. + − / × (÷ circled)

G The problem does not deal with the same number again and again. The problem does not give the big number. (+ circled) − / × ÷

H The problem deals with the same number again and again. The problem does not give the big number. + − / (× circled) ÷

I The problem does not deal with the same number again and again. The problem gives the big number. + (− circled) / × ÷

J The problem deals with the same number again and again. The problem gives the big number. + − / × (÷ circled)

8

4|24 = 6	9|45 = 5	6|48 = 8	7|42 = 6	3|9 = 3	6|24 = 4	7|49 = 7
8|64 = 8	6|36 = 6	4|12 = 3	8|48 = 6	9|36 = 4	7|14 = 2	6|42 = 7
4|28 = 7	7|28 = 4	4|36 = 9	2|16 = 8	6|18 = 3	5|15 = 3	4|8 = 2
8|72 = 9	7|35 = 5	9|18 = 2	6|12 = 2	7|56 = 8	4|16 = 4	5|45 = 9
9|27 = 3	6|54 = 9	4|32 = 8	7|63 = 9	6|30 = 5	8|56 = 7	3|12 = 4
4|20 = 5	3|15 = 5	5|30 = 6	2|18 = 9	7|21 = 3		

9

9|63 = 7	4|16 = 4	9|54 = 6	9|72 = 8	9|63 = 7	6|42 = 7	9|54 = 6
6|54 = 9	6|36 = 6	4|12 = 3	9|54 = 6	4|20 = 5	9|63 = 7	6|48 = 8

Lesson 51 (continued)

10

Write the facts with no remainders.

A 4|21 — 4|20 = 5

B 4|11 — 4|8 = 2

C 4|19 — 4|16 = 4

D 4|6 — 4|4 = 1

E 4|15 — 4|12 = 3

F 4|3 — 4|0 = 0

11

A 15|705 = 47
−60
105
−105
0

B 6|7324 = 1220
−6
13
−12
12
−12
4

C 26|1348 = 51
−130
48
−26
22

D 8|8408 = 1051
−8
40
−40
8
−8
0

E 21|1542 = 73
−147
72
−63
9

F 27|1234 = 45
−108
154
−135
19

G 43|1824 = 42
−172
104
−86
18

H 37|1268 = 34
−111
158
−148
10

Lesson 52

Facts + Problems + Bonus = TOTAL

1

A 9|81 **B** 9|63 **C** 9|54 **D** 9|72

2

A 9|54 (6) — 9|54 = 6 — 6|54 = 9

B 9|63 (7) — 9|63 = 7 — 7|63 = 9

C 9|72 (8) — 9|72 = 8 — 8|72 = 9

D 9|81 (9) — 9|81 = 9

E 9|90 (10) — 9|90 = 10 — 10|90 = 9

3

A

Box: 6|45

64|4496 = 70
−448
16

B

Box: 7|36

71|3598 = 50
−355
48

C

Boxes: 4|10, 4|13

43|987 = 22
−86
127
−86
41

D

Boxes: 5|5, 5|22

52|5415 = 104
−52
215
−208
7

4

A The problem deals with the same number again and again. The problem does not give the big number. + − / (× circled) ÷

B The problem does not deal with the same number again and again. The problem does not give the big number. (+ circled) − / × ÷

Part 4 continues on the next page.

Lesson 52 (continued)

C The problem deals with the same number again and again. The problem gives the big number. + − × (÷)

D The problem deals with the same number again and again. The problem does not give the big number. + − (×) ÷

5

A There are 6 kids on each team. There are 48 kids. How many teams are there?
Answer: **8** teams — + − × (÷)
Work: 6|48 = 8

B Sherry has some leaves. 8 of them are red and 9 of them are yellow. How many leaves does Sherry have?
Answer: **17** leaves — (+) − × ÷
Work: 8 + 9 = 17

C The parade has 17 bands. 6 of the bands are on floats and the rest of them are marching. How many bands are marching?
Answer: **11** bands — + (−) × ÷
Work: 17 − 6 = 11

D Pete did 35 problems on each page. He did 11 pages of problems. How many problems did he do?
Answer: **385** problems — + − (×) ÷
Work: 35 × 11 = 35 + 35 (shifted) = 385

E Betty types 7 pages an hour. She types 56 pages. How many hours did she type?
Answer: **8** hours — + − × (÷)
Work: 7|56 = 8

6

4|36 = 9	8|64 = 8	7|21 = 3	2|4 = 2	7|56 = 8	3|6 = 2	7|35 = 5
6|48 = 8	5|35 = 7	9|36 = 4	6|18 = 3	7|28 = 4	8|48 = 6	4|8 = 2

Part 6 continues on the next page.

Lesson 52 (continued)

7|7 = 1	4|20 = 5	5|20 = 4	8|72 = 9	10|30 = 3	6|24 = 4	3|9 = 3
7|42 = 6	4|24 = 6	8|40 = 5	6|42 = 7	10|60 = 6	9|27 = 3	4|28 = 7
9|45 = 5	6|54 = 9	7|49 = 7	3|12 = 4	7|63 = 9	5|40 = 8	4|12 = 3
4|16 = 4	2|14 = 7	8|56 = 7	2|12 = 6	4|36 = 9		

7

9|63 = 7	6|42 = 7	9|54 = 6	6|60 = 10	9|72 = 8	9|63 = 7	9|45 = 5
9|72 = 8	9|54 = 6	4|8 = 2	6|36 = 6	9|63 = 7	6|48 = 8	9|54 = 6

8

Write the facts with no remainders.

	Problem	Fact
A	4|5	4|4 = 1
B	4|22	4|20 = 5
C	4|13	4|12 = 3
D	4|9	4|8 = 2
E	4|1	4|0 = 0
F	4|17	4|16 = 4
G	6|46	6|42 = 7
H	6|38	6|36 = 6
I	6|64	6|60 = 10
J	6|50	6|48 = 8
K	6|55	6|54 = 9
L	6|43	6|42 = 7

9

A 38|2739 = 72; −266, 79, −76, remainder 3

B 4|4135 = 1033; −4, 13, −12, 15, −12, remainder 3

C 24|607 = 25; −48, 127, −120, remainder 7

D 47|14,386 = 306; −141, 286, −282, remainder 4

E 7|915 = 130; −7, 21, −21, remainder 5

Lesson 53

Facts + Problems + Bonus = TOTAL

1

A 3|30 **B** 3|21 **C** 3|27 **D** 3|18

2

A [6] 3|18; 3|18 = 6; 6|18 = 3

B [7] 3|21; 3|21 = 7; 7|21 = 3

C [8] 3|24; 3|24 = 8; 8|24 = 3

D [9] 3|27; 3|27 = 9; 9|27 = 3

E [10] 3|30; 3|30 = 10; 10|30 = 3

3

A Box: 4|35. 39|3542 = 90; −351, remainder 32

B Boxes: 6|25, 6|37. 62|25,174 = 406; −248, 374, −372, remainder 2

C Boxes: 7|52, 7|8. 73|51,852 = 710; −511, 75, −73, remainder 22

D Box: 6|51. 64|5121 = 80; −512, remainder 1

4

A [7] 9|63; 9|63 = 7; 7|63 = 9

B [10] 9|90; 9|90 = 10; 10|90 = 9

Part 4 continues on the next page.

Lesson 53 (continued)

C [9] 9|81; 9|81 = 9

D [8] 9|72; 9|72 = 8; 8|72 = 9

E [6] 9|54; 9|54 = 6; 6|54 = 9

5

A A woman has 126 plants in her garden. Her daughter has 96. How many more plants does the woman have than her daughter?
Answer: **30** plants — + (−) × ÷
Work: 126 − 96 = 30

B Vic drinks water 6 times every day. After 42 days go by, how many times will Vic have drunk water?
Answer: **252** times — + − (×) ÷
Work: 42 × 6 = 252

C There are 4 cans in each box. There are 24 cans in boxes. How many boxes are there?
Answer: **6** boxes — + − × (÷)
Work: 4|24 = 6

D The doctor saw 32 patients on Monday, 20 patients on Tuesday, 4 patients on Wednesday, and 7 patients on Thursday. How many patients did she see?
Answer: **61** patients — (+) − × ÷
Work: 30 + 20 + 4 + 7 = 61

E Every bag Mike carried weighed 18 kilograms. If he carried 9 bags, how many kilograms did he carry?
Answer: **162** kilograms — + − (×) ÷
Work: 18 × 9 = 162

F Millie went to the bank 9 times every month. In all, she went to the bank 126 times. How many months did Millie go to the bank?
Answer: **14** months — + − × (÷)
Work: 9|126 = 14; −9, 36, −36, 0

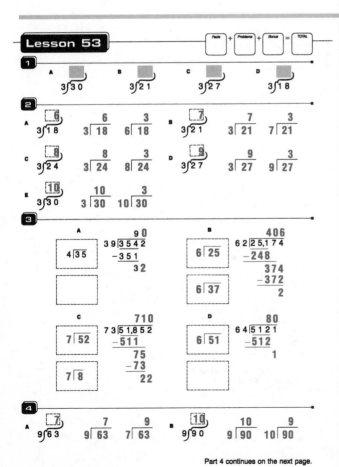

Lesson 53 (continued)

6

8)72 = 9	7)42 = 6	3)12 = 4	7)21 = 3	4)28 = 7	9)36 = 4	5)30 = 6
5)25 = 5	1)6 = 6	7)49 = 7	2)6 = 3	6)18 = 3	7)28 = 4	8)56 = 7
1)2 = 2	7)56 = 8	9)9 = 1	4)0 = 0	8)48 = 6	4)20 = 5	2)10 = 5
6)30 = 5	3)6 = 2	4)12 = 3	5)10 = 2	9)27 = 3	6)6 = 1	7)0 = 0
7)63 = 9	6)42 = 7	2)8 = 4	8)64 = 8	10)40 = 4	7)35 = 5	3)15 = 5
6)54 = 9	4)36 = 9	5)45 = 9	10)10 = 1	5)5 = 1		

7

9)90 = 10	9)63 = 7	9)81 = 9	9)54 = 6	3)18 = 6	9)90 = 10	3)21 = 7
3)18 = 6	9)81 = 9	3)21 = 7	9)63 = 7	3)18 = 6	9)54 = 6	9)81 = 9

8 Write the facts with no remainders.

- A 6)39 — 6)36 = 6
- B 6)58 — 6)54 = 9
- C 6)62 — 6)60 = 10
- D 6)44 — 6)42 = 7
- E 6)40 — 6)36 = 6
- F 6)49 — 6)48 = 8

9

- A 35)2843 = 81; −280; 43; −35; 8
- B 6)2523 = 420; −24; 12; −12; 03
- C 20)310 = 15; −20; 110; −100; 10
- D 48)240 = 5; −240; 0
- E 62)1814 = 29; −124; 574; −558; 16

Lesson 54

Test + Facts + Problems + Bonus = TOTAL

1

- A 3)24
- B 3)18
- C 3)30
- D 3)21

2

- A 3)18 [6]; 3)18 = 6; 6)18 = 3
- B 3)21 [7]; 3)21 = 7; 7)21 = 3
- C 3)24 [8]; 3)24 = 8; 8)24 = 3
- D 3)27 [9]; 3)27 = 9; 9)27 = 3
- E 3)30 [10]; 3)30 = 10; 10)30 = 3

3

- A [8)83] 84)825 = 9; −756; 69
- B [3)31] 33)309 = 9; −297; 12

4

- A 9)72 [8]; 9)72 = 8; 8)72 = 9
- B 9)54 [6]; 9)54 = 6; 6)54 = 9
- C 9)90 [10]; 9)90 = 10; 10)90 = 9
- D 9)81 [9]; 9)81 = 9
- E 9)63 [7]; 9)63 = 7; 7)63 = 9

Lesson 54 (continued)

5

- A 6)3020 = 503; −30; 20; −18; 2
- B 75)3052 = 40; −300; 52
- C 34)3618 = 106; −34; 218; −204; 14
- D 43)2186 = 50; −215; 36
- E 15)452 = 30; −45; 2

6

2)16 = 8	7)28 = 4	4)32 = 8	8)40 = 5	6)48 = 8	5)35 = 7	4)16 = 4
1)8 = 8	9)18 = 2	10)50 = 5	4)24 = 6	7)14 = 2	8)64 = 8	6)24 = 4
6)60 = 10	7)35 = 5	3)9 = 3	2)2 = 1	8)72 = 9	4)8 = 2	6)12 = 2
4)4 = 1	8)48 = 6	2)18 = 9	4)40 = 10	9)45 = 5	6)36 = 6	3)3 = 1
5)40 = 8	10)70 = 7	8)80 = 10	5)15 = 3	1)4 = 4	8)56 = 7	5)50 = 10
10)90 = 9	2)20 = 10	7)49 = 7	5)20 = 4	7)56 = 8		

7

9)72 = 8	3)21 = 7	9)81 = 9	3)18 = 6	9)54 = 6	9)90 = 10	3)21 = 7
9)90 = 10	9)63 = 7	3)18 = 6	9)72 = 8	3)21 = 7	9)81 = 9	9)54 = 6
9)63 = 7	9)90 = 10	9)81 = 9	3)21 = 7			

Lesson 54 (continued)

8 Write the facts with no remainders.

- A 6)52 — 6)48 = 8
- B 6)41 — 6)36 = 6
- C 6)47 — 6)42 = 7
- D 6)37 — 6)36 = 6
- E 6)53 — 6)48 = 8
- F 6)56 — 6)54 = 9

9 Don't forget to circle the right sign.

A Hugo wants to make 42 stacks of boxes. If he puts 126 boxes in every stack, how many boxes will he use?

[5292] boxes   + −   (×) ÷

126 × 42; 252; + 504; 5292

B There are 3 eggs in each nest. There are 99 eggs in nests. How many nests are there?

[33] nests   + −   × (÷)

3)99 = 33

C Bessie's report has 14 pages, Del's report has 22 pages, and Liz's report has 47 pages. How many pages are there in all?

[83] pages   (+) −   × ÷

14 + 22 + 47 = 83

D Airplanes land at Midtown Airport for 6 days. 41 airplanes land each day. How many airplanes land?

[246] airplanes   + −   (×) ÷

41 × 6 = 246

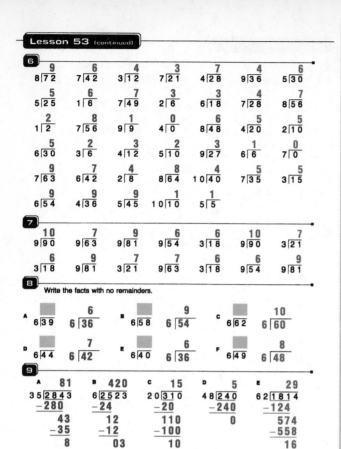

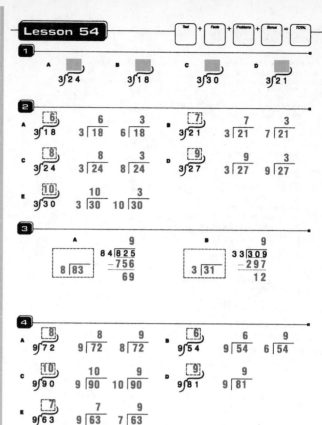

Lesson 55

Facts + Problems + Bonus = TOTAL

1

A 8⟌32 = ☐ B 8⟌16 = ☐ C 8⟌8 = ☐ D 8⟌24 = ☐

2

A 8⟌8 = [1]; 8⟌8 = 1; 1⟌8 = 8
B 8⟌16 = [2]; 8⟌16 = 2; 2⟌16 = 8
C 8⟌24 = [3]; 8⟌24 = 3; 3⟌24 = 8
D 8⟌32 = [4]; 8⟌32 = 4; 4⟌32 = 8

3

A 9⟌63 = 7 B 9⟌81 = 9 C 9⟌72 = 8 D 9⟌54 = 6

4

A 3⟌27 = [9]; 3⟌27 = 9; 9⟌27 = 3
B 3⟌30 = [10]; 3⟌30 = 10; 10⟌30 = 3
C 3⟌24 = [8]; 3⟌24 = 8; 8⟌24 = 3
D 3⟌18 = [6]; 3⟌18 = 6; 6⟌18 = 3
E 3⟌21 = [7]; 3⟌21 = 7; 7⟌21 = 3

5

A [4⟌43] 43⟌428 = 9; −387; 41
B [8⟌82] 84⟌815 = 9; −756; 59
C [6⟌61] 62⟌612 = 9; −558; 54

Lesson 55 (continued)

6

4⟌20 = 5 6⟌42 = 7 10⟌80 = 8 9⟌81 = 9 9⟌63 = 7 2⟌14 = 7 5⟌0 = 0
6⟌6 = 1 7⟌63 = 9 9⟌27 = 3 8⟌48 = 6 4⟌12 = 3 10⟌20 = 2 3⟌15 = 5
8⟌64 = 8 9⟌72 = 8 4⟌0 = 0 3⟌6 = 2 4⟌28 = 7 9⟌36 = 4 2⟌12 = 6
5⟌30 = 6 7⟌70 = 10 6⟌18 = 3 4⟌36 = 9 8⟌56 = 7 5⟌45 = 9 9⟌90 = 10
9⟌54 = 6 8⟌72 = 9 7⟌21 = 3 3⟌12 = 4 1⟌7 = 7 2⟌4 = 2 10⟌0 = 0
5⟌25 = 5 9⟌0 = 0 6⟌54 = 9 7⟌42 = 6 6⟌30 = 5

7

3⟌21 = 7 3⟌30 = 10 3⟌24 = 8 8⟌16 = 2 3⟌18 = 6 8⟌24 = 3 3⟌24 = 8
8⟌16 = 2 3⟌18 = 6 3⟌30 = 10 3⟌27 = 9 8⟌24 = 3 8⟌16 = 2 3⟌30 = 10
3⟌27 = 9 3⟌24 = 8 8⟌24 = 3 3⟌21 = 7 3⟌27 = 9

8

Don't forget to circle the right sign.

A Mack has 28 boards. 19 of the boards are long. The rest of the boards are short. How many short boards does Mack have?

[9] short boards + (−) × ÷

28 − 19 = 9

Part 8 continues on the next page.

Lesson 55 (continued)

B Jan rides her bike to school 3 days a week. She rides it to school 141 times. How many weeks did Jan ride her bike to school?

[47] weeks + − × (÷)

3⟌141 = 47; −12; 21; −21; 0

C Every time Sara goes to the creek, she gets 8 rocks. She goes to the creek 34 times. How many rocks does she get?

[272] rocks + − (×) ÷

34 × 8 = 272

D Lisa had 347 nuts. Then she found some more. Now she has 872. How many did she find?

[525] nuts + (−) × ÷

872 − 347 = 525

9

A 22⟌6714 = 305; −66; 114; −110; 4
B 58⟌2341 = 40; −232; 21
C 26⟌784 = 30; −78; 4
D 7⟌8423 = 1203; −7; 14; −14; 23; −21; 2
E 24⟌789 = 32; −72; 69; −48; 21
F 8⟌1000 = 125; −8; 20; −16; 40; −40; 0

Lesson 56

Facts + Problems + Bonus = TOTAL

1

A 8⟌16 = ☐ B 8⟌32 = ☐ C 8⟌24 = ☐ D 8⟌8 = ☐

2

A 8⟌8 = [1]; 8⟌8 = 1; 1⟌8 = 8
B 8⟌16 = [2]; 8⟌16 = 2; 2⟌16 = 8
C 8⟌24 = [3]; 8⟌24 = 3; 3⟌24 = 8
D 8⟌32 = [4]; 8⟌32 = 4; 4⟌32 = 8

3

A 74⟌758 = 10; −74; 18
B 82⟌816 = 9; −738; 78
C 53⟌528 = 9; −477; 51

4

A 3⟌24 = [8]; 3⟌24 = 8; 8⟌24 = 3
B 3⟌18 = [6]; 3⟌18 = 6; 6⟌18 = 3
C 3⟌30 = [10]; 3⟌30 = 10; 10⟌30 = 3
D 3⟌21 = [7]; 3⟌21 = 7; 7⟌21 = 3
E 3⟌27 = [9]; 3⟌27 = 9; 9⟌27 = 3

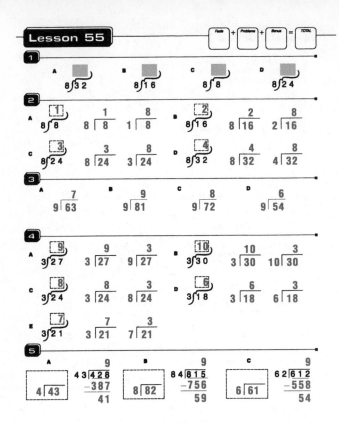

Lesson 56 (continued)

5

9)45 = 5	8)56 = 7	4)24 = 6	7)28 = 4	5)40 = 8	7)14 = 2	9)54 = 6
8)64 = 8	9)0 = 0	7)7 = 1	2)6 = 3	9)90 = 10	8)40 = 5	3)3 = 1
7)35 = 5	8)48 = 6	4)16 = 4	3)9 = 3	10)30 = 3	9)18 = 2	2)8 = 4
4)8 = 2	6)0 = 0	9)63 = 7	2)10 = 5	7)49 = 7	6)12 = 2	3)0 = 0
2)0 = 0	10)100 = 10	9)81 = 9	5)10 = 2	6)48 = 8	9)72 = 8	4)40 = 10
8)72 = 9	6)36 = 6	4)32 = 8	7)56 = 8	6)24 = 4		

6

3)21 = 7	3)30 = 10	3)24 = 8	8)16 = 2	3)27 = 9	8)24 = 3	3)30 = 10
3)27 = 9	3)24 = 8	3)18 = 6	8)24 = 3	3)30 = 10	3)21 = 7	8)16 = 2
3)18 = 6	3)27 = 9	3)24 = 8	8)16 = 2	8)24 = 3		

7

Write the facts with no remainders.

A 9)57 — 9)54 = 6
B 9)76 — 9)72 = 8
C 9)83 — 9)81 = 9
D 9)68 — 9)63 = 7
E 9)59 — 9)54 = 6
F 9)87 — 9)81 = 9

Lesson 56 (continued)

8

A Gus made dinner 3 times each week. He made dinner for 28 weeks. How many times did he make dinner?

84 times — 28 × 3 = 84

B There are 24 girls at Friendly School. There are 21 boys and 6 teachers. How many people are at Friendly School?

51 people — 24 + 21 + 6 = 51

C Mr. Bear hooked up 62 telephones every month. He hooked up 434 telephones. How many months did Mr. Bear hook up telephones?

7 months — 62)434 = 7; −434; 0

D The town has 37 stores. 14 of them are closed on Saturdays. The rest are open. How many stores are open?

23 stores — 37 − 14 = 23

E Every year Jackie got 41 letters. She got 369 letters. How many years went by?

9 years — 41)369 = 9; −369; 0

F Marjie chewed 8 pieces of celery every week. She chewed 64 pieces of celery. How many weeks did Marjie chew celery?

8 weeks — 8)64 = 8

G Patty dug 5 holes each hour. She dug holes for 230 hours. How many holes did Patty dig?

1150 holes — 230 × 5 = 1150

Lesson 56 (continued)

9

```
A  33)2003 = 60      B  65)2608 = 40      C  80)3421 = 42
     -198                 -260                 -320
       23                    8                  221
                                               -160
                                                 61

D  8)5764 = 720      E  35)1191 = 34      F  24)1315 = 54
    -56                   -105                 -120
     16                    141                  115
    -16                   -140                  -96
      4                      1                   19

G  50)3100 = 62      H  52)5178 = 99      I  37)958 = 25
     -300                 -468                 -74
      100                  498                 218
     -100                 -468                -185
        0                   30                  33
```


Lesson 57

Test + Facts + Problems + Bonus = TOTAL

1

A 8)24 → [3]; 8)24 = 3; 3)24 = 8
B 8)16 → [2]; 8)16 = 2; 2)16 = 8
C 8)32 → [4]; 8)32 = 4; 4)32 = 8
D 8)8 → [1]; 8)8 = 1; 1)8 = 8

2

A 3)27 = 9 B 3)18 = 6 C 3)24 = 8 D 3)21 = 7

3

3)21 = 7	2)18 = 9	7)42 = 6	9)81 = 9	4)28 = 7	8)56 = 7	10)60 = 6
6)30 = 5	8)72 = 9	2)16 = 8	3)27 = 9	7)21 = 3	9)36 = 4	4)36 = 9
2)4 = 2	3)24 = 8	9)27 = 3	4)12 = 3	3)6 = 2	7)0 = 0	3)30 = 10
9)9 = 1	3)12 = 4	8)48 = 6	7)70 = 10	2)20 = 10	9)63 = 7	4)20 = 5
6)18 = 3	2)2 = 1	9)72 = 8	7)63 = 9	6)42 = 7	5)20 = 4	9)90 = 10
6)54 = 9	3)18 = 6	9)54 = 6	8)64 = 8	3)15 = 5		

4

3)18 = 6	8)32 = 4	8)16 = 2	3)24 = 8	8)32 = 4	9)63 = 7	3)27 = 9
9)81 = 9	8)16 = 2	3)21 = 7	8)32 = 4	9)54 = 6	8)24 = 3	9)72 = 8

5

A. Jimmy has toys. Ten are yo-yos and 15 are trucks. How many toys does he have?

[25] toys

$$\begin{array}{r} 10 \\ + 15 \\ \hline 25 \end{array}$$

B. Gussie bought oil 13 times every year. She bought oil 65 times. How many years did Gussie buy oil?

[5] years

$$\begin{array}{r} 5 \\ 13\overline{)65} \\ -65 \\ \hline 0 \end{array}$$

C. Mr. Eagle used 3 eggs each time he baked bread. In all, he used 18 eggs to make bread. How many times did he bake bread?

[6] times

$$\begin{array}{r} 6 \\ 3\overline{)18} \end{array}$$

D. Every year Lucy went to the park 16 times. She went for 42 years. How many times did Lucy go to the park?

[672] times

$$\begin{array}{r} 16 \\ \times 42 \\ \hline 32 \\ 64 \\ \hline 672 \end{array}$$

E. Ming baby-sat 7 times each month. He baby-sat 168 times. How many months did Ming baby-sit?

[24] months

$$\begin{array}{r} 24 \\ 7\overline{)168} \\ -14 \\ \hline 28 \\ -28 \\ \hline 0 \end{array}$$

F. Marty tested 27 radios each day. He tested 189 radios. How many days did Marty test radios?

[7] days

$$\begin{array}{r} 7 \\ 27\overline{)189} \\ -189 \\ \hline 0 \end{array}$$

6

A.
$$\begin{array}{r} 97 \\ 31\overline{)3024} \\ -279 \\ \hline 234 \\ -217 \\ \hline 17 \end{array}$$

B.
$$\begin{array}{r} 409 \\ 21\overline{)8601} \\ -84 \\ \hline 201 \\ -189 \\ \hline 12 \end{array}$$

C.
$$\begin{array}{r} 50 \\ 46\overline{)2304} \\ -230 \\ \hline 4 \end{array}$$

D.
$$\begin{array}{r} 250 \\ 4\overline{)1001} \\ -8 \\ \hline 20 \\ -20 \\ \hline 1 \end{array}$$

E.
$$\begin{array}{r} 1013 \\ 7\overline{)7091} \\ -7 \\ \hline 09 \\ -7 \\ \hline 21 \\ -21 \\ \hline 0 \end{array}$$

F.
$$\begin{array}{r} 96 \\ 24\overline{)2318} \\ -216 \\ \hline 158 \\ -144 \\ \hline 14 \end{array}$$

G.
$$\begin{array}{r} 22 \\ 43\overline{)987} \\ -86 \\ \hline 127 \\ -86 \\ \hline 41 \end{array}$$

7 Write the facts with no remainders.

A. ▨ 9$\overline{)93}$ $\dfrac{10}{9\overline{)90}}$

B. ▨ 9$\overline{)85}$ $\dfrac{9}{9\overline{)81}}$

C. ▨ 9$\overline{)56}$ $\dfrac{6}{9\overline{)54}}$

D. ▨ 9$\overline{)77}$ $\dfrac{8}{9\overline{)72}}$

E. ▨ 9$\overline{)70}$ $\dfrac{7}{9\overline{)63}}$

F. ▨ 9$\overline{)61}$ $\dfrac{6}{9\overline{)54}}$

G. ▨ 9$\overline{)50}$ $\dfrac{5}{9\overline{)45}}$

H. ▨ 9$\overline{)97}$ $\dfrac{10}{9\overline{)90}}$

I. ▨ 9$\overline{)88}$ $\dfrac{9}{9\overline{)81}}$

J. ▨ 9$\overline{)38}$ $\dfrac{4}{9\overline{)36}}$

K. ▨ 9$\overline{)75}$ $\dfrac{8}{9\overline{)72}}$

L. ▨ 9$\overline{)69}$ $\dfrac{7}{9\overline{)63}}$

Lesson 58 [Facts] + [Problems] + [Bonus] = [TOTAL]

1

A. $\dfrac{[3]}{8\overline{)24}}$ $\dfrac{3}{8\overline{)24}}$ $\dfrac{8}{3\overline{)24}}$

B. $\dfrac{[1]}{8\overline{)8}}$ $\dfrac{1}{8\overline{)8}}$ $\dfrac{8}{1\overline{)8}}$

C. $\dfrac{[4]}{8\overline{)32}}$ $\dfrac{4}{8\overline{)32}}$ $\dfrac{8}{4\overline{)32}}$

D. $\dfrac{[2]}{8\overline{)16}}$ $\dfrac{2}{8\overline{)16}}$ $\dfrac{8}{2\overline{)16}}$

2

$\dfrac{8}{9\overline{)72}}$ $\dfrac{6}{3\overline{)18}}$ $\dfrac{6}{6\overline{)36}}$ $\dfrac{5}{7\overline{)35}}$ $\dfrac{9}{8\overline{)72}}$ $\dfrac{5}{9\overline{)45}}$ $\dfrac{7}{4\overline{)28}}$

$\dfrac{8}{4\overline{)32}}$ $\dfrac{7}{9\overline{)63}}$ $\dfrac{8}{7\overline{)56}}$ $\dfrac{8}{8\overline{)64}}$ $\dfrac{3}{3\overline{)9}}$ $\dfrac{7}{2\overline{)14}}$ $\dfrac{8}{6\overline{)48}}$

$\dfrac{7}{3\overline{)21}}$ $\dfrac{7}{8\overline{)56}}$ $\dfrac{1}{4\overline{)4}}$ $\dfrac{2}{9\overline{)18}}$ $\dfrac{6}{4\overline{)24}}$ $\dfrac{1}{7\overline{)7}}$ $\dfrac{10}{8\overline{)80}}$

$\dfrac{4}{4\overline{)16}}$ $\dfrac{9}{9\overline{)81}}$ $\dfrac{5}{8\overline{)40}}$ $\dfrac{4}{7\overline{)28}}$ $\dfrac{4}{10\overline{)40}}$ $\dfrac{1}{3\overline{)3}}$ $\dfrac{6}{2\overline{)12}}$

$\dfrac{1}{6\overline{)6}}$ $\dfrac{6}{8\overline{)48}}$ $\dfrac{5}{3\overline{)15}}$ $\dfrac{4}{6\overline{)24}}$ $\dfrac{2}{7\overline{)14}}$ $\dfrac{1}{5\overline{)5}}$ $\dfrac{8}{3\overline{)24}}$

$\dfrac{9}{3\overline{)27}}$ $\dfrac{4}{3\overline{)12}}$ $\dfrac{6}{9\overline{)54}}$ $\dfrac{7}{7\overline{)49}}$ $\dfrac{7}{5\overline{)35}}$

3

$\dfrac{4}{8\overline{)32}}$ $\dfrac{6}{9\overline{)54}}$ $\dfrac{10}{3\overline{)30}}$ $\dfrac{2}{8\overline{)16}}$ $\dfrac{7}{3\overline{)21}}$ $\dfrac{4}{8\overline{)32}}$ $\dfrac{8}{9\overline{)72}}$

$\dfrac{7}{9\overline{)63}}$ $\dfrac{4}{8\overline{)32}}$ $\dfrac{6}{3\overline{)18}}$ $\dfrac{9}{9\overline{)81}}$ $\dfrac{3}{8\overline{)24}}$ $\dfrac{9}{3\overline{)27}}$ $\dfrac{2}{8\overline{)16}}$

4

A. A swimmer swam 8 kilometers every hour. She swam 96 kilometers. How many hours did she swim?

[12] hours

$$\begin{array}{r} 12 \\ 8\overline{)96} \\ -8 \\ \hline 16 \\ -16 \\ \hline 0 \end{array}$$

Part 4 continues on the next page.

B. The train made 3 stops every day it ran. It made 72 stops. How many days did the train run?

[24] days

$$\begin{array}{r} 24 \\ 3\overline{)72} \\ -6 \\ \hline 12 \\ -12 \\ \hline 0 \end{array}$$

C. Maria has 155 pens. She keeps 42 pigs in each pen. How many pigs does she have?

[6510] pigs

$$\begin{array}{r} 155 \\ \times 42 \\ \hline 310 \\ 620 \\ \hline 6510 \end{array}$$

D. The mover lifted 23 boxes every hour. He lifted 138 boxes. How many hours did he lift boxes?

[6] hours

$$\begin{array}{r} 6 \\ 23\overline{)138} \\ -138 \\ \hline 0 \end{array}$$

E. Carol wants to build 58 houses. She needs to hire 12 people to work on each house. How many people does she need to hire?

[696] people

$$\begin{array}{r} 58 \\ \times 12 \\ \hline 116 \\ 58 \\ \hline 696 \end{array}$$

F. Toni grew 15 carrots in 2002. She grew 11 carrots in 2004. How many more carrots did she grow in 2002 than in 2004?

[4] carrots

$$\begin{array}{r} 15 \\ -11 \\ \hline 4 \end{array}$$

5

A.
$$\begin{array}{r} 8 \\ 42\overline{)376} \\ -336 \\ \hline 40 \end{array}$$

B.
$$\begin{array}{r} 108 \\ 87\overline{)9481} \\ -87 \\ \hline 781 \\ -696 \\ \hline 85 \end{array}$$

C.
$$\begin{array}{r} 80 \\ 36\overline{)2882} \\ -288 \\ \hline 2 \end{array}$$

D.
$$\begin{array}{r} 76 \\ 13\overline{)1000} \\ -91 \\ \hline 90 \\ -78 \\ \hline 12 \end{array}$$

Part 5 continues on the next page.

E
100
99)9909
−99
09

F
12
54)678
−54
138
−108
30

G
301
7)2111
−21
11
−7
4

6

Write the facts with no remainders.

A	B	C
9)74 → 9)72 = 8	9)96 → 9)90 = 10	9)60 → 9)54 = 6

D	E	F
9)88 → 9)81 = 9	9)67 → 9)63 = 7	9)84 → 9)81 = 9

G	H	I
3)31 → 3)30 = 10	3)23 → 3)21 = 7	3)20 → 3)18 = 6

J	K	L
3)25 → 3)24 = 8	3)22 → 3)21 = 7	3)28 → 3)27 = 9

M	N	O
9)57 → 9)54 = 6	9)78 → 9)72 = 8	9)64 → 9)63 = 7

Lesson 59

Facts + Problems + Bonus = TOTAL

1

A	B	C	D
8)24 = 3	8)32 = 4	8)8 = 1	8)16 = 2

2

8)32 = 4	7)42 = 6	10)70 = 7	9)63 = 7	8)8 = 1	8)24 = 3	6)48 = 8
4)32 = 8	5)45 = 9	2)18 = 9	6)54 = 9	4)28 = 7	8)64 = 8	7)49 = 7
2)16 = 8	6)36 = 6	8)8 = 1	3)24 = 8	7)56 = 8	9)81 = 9	5)35 = 7
8)56 = 7	5)40 = 8	4)24 = 6	9)54 = 6	3)27 = 9	8)16 = 2	7)63 = 9
10)60 = 6	7)35 = 5	2)14 = 7	8)48 = 6	3)18 = 6	6)42 = 7	5)30 = 6
3)21 = 7	9)72 = 8	8)40 = 5	4)36 = 9	8)72 = 9		

3

A
96
42)4035
−378
255
−252
3

B
22
31)684
−62
64
−62
2

C
140
7)981
−7
28
−28
1

D
900
9)8106
−81
06

E
704
4)2818
−28
18
−16
2

F
75
67)5034
−469
344
−335
9

4

A Dennis drank 2 glasses of milk every time he ate lunch. He drank 24 glasses of milk at lunches. How many lunches did he have?

2)24 = 12; −2, 4, −4, 0

12 lunches

B Band A has 4 drum players, band B has 2 drum players, and band C has 11 drum players. How many drum players are there?

4 + 2 + 11 = 17

17 drum players

C Jenny milked 26 goats every night. She milked 78 goats. How many nights did Jenny milk goats?

26)78 = 3; −78, 0

3 nights

D Faye grows 4 centimeters each year. After 8 years go by, how many centimeters will she have grown?

4 × 8 = 32

32 centimeters

E A man must find 482 people to work for him. He finds 193 people the first day. How many more people must he find?

482 − 193 = 289

289 people

F Lonnie bought 3 liters of milk whenever he went to the store. He bought 21 liters of milk. How many times did he go to the store?

3)21 = 7

7 times

5

Write the facts with no remainders.

A	B	C
3)26 → 3)24 = 8	3)29 → 3)27 = 9	3)19 → 3)18 = 6

D	E	F
3)32 → 3)30 = 10	3)22 → 3)21 = 7	3)20 → 3)18 = 6

Lesson 60

Test + Facts + Problems + Bonus = TOTAL

1

A Every square has 4 sides. There are 572 squares. How many sides are there?

572 × 4 = 2288

2288 sides

B Angel has 163 friends. 44 are girls and the rest are boys. How many of Angel's friends are boys?

163 − 44 = 119

119 boys

C Shannon took 2 breaks each day that she worked. She worked for 18 days. How many breaks did Shannon take?

18 × 2 = 36

36 breaks

D Rick has 24 pieces of pipe. He gets 17 pieces. How many pieces of pipe does Rick have now?

24 + 17 = 41

41 pieces

E Carmen painted 16 pictures every month. She painted 64 pictures. How many months did Carmen paint pictures?

16)64 = 4; −64, 0

4 months

F Gerald walked 9 blocks to work every day. He walked 90 blocks. How many days did he go to work?

9)90 = 10

10 days

2

9)54 = 6	7)49 = 7	3)18 = 6	4)16 = 4	2)14 = 7	8)24 = 3	2)6 = 3
5)25 = 5	8)8 = 1	7)63 = 9	3)27 = 9	9)36 = 4	2)18 = 9	8)32 = 4

Part 2 continues on the next page.

4)32 = 8	7)56 = 8	4)12 = 3	7)42 = 6	8)16 = 2	5)35 = 7	9)81 = 9
3)30 = 10	5)20 = 4	9)63 = 7	3)24 = 8	2)16 = 8	4)24 = 6	2)12 = 6
5)15 = 3	4)8 = 2	2)10 = 5	4)36 = 9	9)18 = 2	9)72 = 8	1)4 = 4
1)7 = 7	4)20 = 5	9)27 = 3	9)90 = 10	4)28 = 7		

3

```
A      45        B       211       C      145
    7)321          19)4017          4)581
     -28            -38              -4
      41             21               18
     -35            -19              -16
       6             27                21
                    -19               -20
                      8                 1
```

```
D     185        E       344       F       111
    6)1110         11)3789          81)9004
     -6             -33              -81
      51             48               90
     -48            -44              -81
      30             49               94
     -30            -44              -81
       0              5               13
```

Part 3 continues on the next page.


```
G      40        H       34        I       42
   65)2608         35)1191          80)3421
     -260            -105             -320
        8             141              221
                     -140             -160
                        1               61
```

4

Write the facts with no remainders.

A 3)25 → 3)24 = 8	B 3)19 → 3)18 = 6	C 3)31 → 3)30 = 10
D 3)23 → 3)21 = 7	E 3)32 → 3)30 = 10	F 3)28 → 3)27 = 9
G 8)4 → 8)0 = 0	H 8)20 → 8)16 = 2	I 8)36 → 8)32 = 4
J 8)13 → 8)8 = 1	K 8)34 → 8)32 = 4	L 8)23 → 8)16 = 2

Lesson 61

Facts + Problems + Bonus = TOTAL

1

6)36 = 6	9)63 = 7	4)16 = 4	2)16 = 8	3)21 = 7	6)12 = 2	2)14 = 7
5)35 = 7	7)56 = 8	6)54 = 9	9)72 = 8	7)21 = 3	4)20 = 5	6)42 = 7
6)24 = 4	3)24 = 8	8)48 = 6	6)18 = 3	10)40 = 4	9)54 = 6	7)49 = 7
6)48 = 8	3)9 = 3	3)27 = 9	7)63 = 9	7)42 = 6	8)16 = 2	4)8 = 2
3)18 = 6	4)40 = 10	9)81 = 9	2)6 = 3	3)30 = 10	10)100 = 10	9)18 = 2
8)32 = 4	4)12 = 3	8)24 = 3	6)30 = 5	5)45 = 9		

2

A June saw 17 birds in one tree and 14 birds in another tree. She saw 22 birds on the ground. How many birds did June see?

53 birds

```
  17
  14
+ 22
  53
```

B Alice's company built 13 houses every week. They built 494 houses. How many weeks did they build houses?

38 weeks

```
     38
13)494
  -39
   104
  -104
     0
```

C All triangles have 3 sides. Carlos drew 378 triangles. How many sides did he draw?

1134 sides

```
 378
×  3
1134
```

D There were 1140 sailboats sailing on lakes. There were 76 sailboats on every lake. How many lakes had sailboats?

15 lakes

```
      15
76)1140
   -76
    380
   -380
```

Part 2 continues on the next page.

E All motorcycles have 2 wheels. There are 478 motorcycles. How many wheels are there?

956 wheels

```
 478
×  2
 956
```

3

```
A    138      B    578      C   3409     D    72      E   1207
 36)4992       17)9841        2)6819       9)648        8)9656
   -36           -85          -6           -63          -8
   139           134           8            18           16
  -108          -119          -8           -18          -16
   312           151           19            0            56
  -288          -136          -18                        -56
    24            15            1                           0
```

```
F     98      G     99      H     23     I   306      J    151
 23)2264       34)3382       58)1357       3)918       26)3948
   -207          -306          -116         -9           -26
    194           322           197          18           134
   -184          -306          -174         -18          -130
     10            16            23           0            48
                                                          -26
                                                           22
```

4

Write the facts with no remainders.

A 8)35 → 8)32 = 4	B 8)12 → 8)8 = 1	C 8)28 → 8)24 = 3
D 8)6 → 8)0 = 0	E 8)21 → 8)16 = 2	F 8)15 → 8)8 = 1

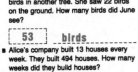

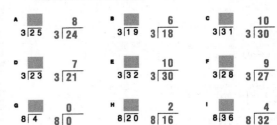

Lesson 62

Facts + Problems + Bonus = TOTAL

1

9⟌72 = 8	5⟌10 = 2	4⟌36 = 9	3⟌12 = 4	8⟌16 = 2	5⟌30 = 6	4⟌16 = 4
4⟌20 = 5	9⟌81 = 9	3⟌9 = 3	6⟌24 = 4	5⟌25 = 5	7⟌63 = 9	9⟌63 = 7
3⟌27 = 9	5⟌15 = 3	8⟌24 = 3	6⟌12 = 2	7⟌42 = 6	3⟌15 = 5	5⟌35 = 7
6⟌18 = 3	3⟌6 = 2	4⟌24 = 6	9⟌54 = 6	5⟌45 = 9	6⟌30 = 5	4⟌32 = 8
8⟌32 = 4	4⟌8 = 2	3⟌24 = 8	5⟌20 = 4	8⟌8 = 1	3⟌18 = 6	5⟌40 = 8
6⟌42 = 7	4⟌28 = 7	6⟌36 = 6	3⟌21 = 7	4⟌12 = 3		

2

A Each bowl has 78 pieces of popcorn in it. If there are 2028 pieces of popcorn in bowls, how many bowls are there?

26 bowls

```
     26
78⟌2028
  -156
   468
  -468
     0
```

B There are 8 bottles in each case. There are 64 cases. How many bottles are there?

512 bottles

```
  64
× 8
 512
```

C I have 756 pieces of cake. Every cake has 9 pieces. How many cakes do I have?

84 cakes

```
   84
9⟌756
 -72
  36
 -36
   0
```

D Amy drove 214 kilometers. Lola drove 108 kilometers. How many more kilometers did Amy drive than Lola?

106 kilometers

```
 214
-108
 106
```

Part 2 continues on the next page.

E Nathan has 476 coins. Roy has 58 more coins than Nathan. How many coins does Roy have?

534 coins

```
 476
+ 58
 534
```

F Heather buys 486 pieces of wood. She wants to make tables that use 81 pieces each. How many tables can she make?

6 tables

```
     6
81⟌486
  -486
     0
```

G There are 24 girls at Clarksville School. There are 21 boys and 6 teachers. How many people are at Clarksville School?

51 people

```
 24
 21
+ 6
 51
```

H The shopping center has 37 stores. 14 of them are closed on Sundays. The rest are open. How many stores are open?

23 stores

```
 37
-14
 23
```

I Every year Nina went to the park 16 times. She went for 42 years. How many times did Nina go to the park?

672 times

```
  16
× 42
  32
 64
 672
```

3

A
```
  123
8⟌987
 -8
  18
 -16
   27
  -24
    3
```

B
```
    74
58⟌4293
  -406
    233
   -232
      1
```

C
```
     729
74⟌53,956
  -518
    215
   -148
     676
    -666
      10
```

D
```
    125
11⟌1385
  -11
    28
   -22
     65
    -55
     10
```

E
```
   86
9⟌780
 -72
   60
  -54
    6
```

Part 3 continues on the next page.

F
```
  1598
5⟌7993
 -5
  29
 -25
   49
  -45
    43
   -40
     3
```

G
```
    96
74⟌7148
  -666
    488
   -444
     44
```

H
```
   956
3⟌2869
 -27
   16
  -15
    19
   -18
     1
```

I
```
    98
89⟌8740
  -801
    730
   -712
     18
```

J
```
   469
5⟌2345
 -20
   34
  -30
    45
   -45
     0
```

4

Write the facts with no remainders.

A 8⟌19 → 8⟌16 = 2

B 8⟌10 → 8⟌8 = 1

C 8⟌38 → 8⟌32 = 4

D 8⟌26 → 8⟌24 = 3

E 8⟌17 → 8⟌16 = 2

F 8⟌31 → 8⟌24 = 3

Lesson 63

Facts + Problems + Bonus = TOTAL

1

2⟌6 = 3	5⟌30 = 6	4⟌20 = 5	3⟌12 = 4	5⟌10 = 2	2⟌12 = 6	7⟌49 = 7
3⟌9 = 3	2⟌18 = 9	5⟌15 = 3	3⟌3 = 1	4⟌12 = 3	6⟌42 = 7	4⟌32 = 8
4⟌16 = 4	6⟌36 = 6	3⟌6 = 2	2⟌10 = 5	4⟌8 = 2	6⟌18 = 3	3⟌21 = 7
4⟌36 = 9	2⟌8 = 4	3⟌18 = 6	8⟌64 = 8	4⟌28 = 7	5⟌25 = 5	7⟌56 = 8
3⟌15 = 5	9⟌72 = 8	6⟌12 = 2	4⟌24 = 6	7⟌63 = 9	2⟌14 = 7	5⟌45 = 9
6⟌24 = 4	3⟌27 = 9	6⟌30 = 5	2⟌16 = 8	3⟌24 = 8		

2

A
```
     91
13⟌1185
  -117
     15
    -13
      2
```

B
```
   5051
8⟌40,412
 -40
   41
  -40
    12
    -8
     4
```

C
```
   135
5⟌678
 -5
  17
 -15
   28
  -25
    3
```

D
```
    12
62⟌791
  -62
   171
  -124
    47
```

E
```
    340
6⟌2044
 -18
   24
  -24
    4
```

F
```
     15
90⟌1400
   -90
   500
  -450
    50
```

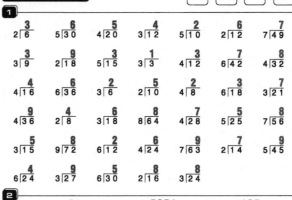

3

A. There are 8 apartments in each building. There are 96 apartments. How many buildings are there?

[12] buildings

```
      12
  8 | 96
    - 8
      16
    - 16
       0
```

B. There are 1269 houses in my hometown. There are 27 houses on every block. How many blocks are in my hometown?

[47] blocks

```
         47
  27 | 1269
     - 108
       189
     - 189
         0
```

C. Janet has 37 carrots. She cooks 20 for dinner. How many carrots does she have left?

[17] carrots

```
    37
  - 20
    17
```

D. Kelly learned 33 new words every month. She learned 1980 new words. How many months did Kelly learn new words?

[60] months

```
         60
  33 | 1980
     - 198
         0
```

E. Ray put 7 glasses on each tray. How many glasses did he put on trays if there are 385 trays?

[2695] glasses

```
     385
   ×   7
    2695
```

F. A farmer had 1060 cows. 7 ran away. How many remain?

[1053] cows

```
    1060
   -   7
    1053
```

4

A.
```
        136
  44 | 5990
     - 44
       159
     - 132
       270
     - 264
         6
```

B.
```
       126
   7 | 883
     - 7
      18
    - 14
      43
    - 42
       1
```

C.
```
         90
  24 | 2165
     - 216
         5
```

1

```
    3        5        2        7        6        8        9
8 | 24   7 | 35   9 | 18   5 | 35   6 | 36   7 | 56   8 | 72

    7        4        8        4        4        5        3
6 | 42   5 | 20   4 | 32   7 | 28   8 | 32   9 | 45   6 | 18

    8        7        2        4        6        4        3
9 | 72   7 | 49   8 | 16   9 | 36   5 | 30   6 | 24   7 | 21

    9        3        5       10        6        2        7
6 | 54   9 | 27   5 | 25   4 | 40   8 | 48   6 | 12   9 | 63

    6        8        5        8        9        5        4
7 | 42   6 | 48   8 | 40   5 | 40   7 | 63   6 | 30   4 | 16

    6        6        7        2        8
9 | 54   4 | 24   8 | 56   7 | 14   8 | 64
```

2

A. A train made 8 trips every month. It made 888 trips. How many months did the train make trips?

[111] months

```
        111
  8 | 888
    - 8
      8
    - 8
      8
    - 8
      0
```

B. Yesterday our dog rolled over 14 times. Today it rolled over 35 times. How many times did our dog roll over in all?

[49] times

```
    14
  + 35
    49
```

C. There are 32 people in each building. There are 187 buildings. How many people are there?

[5984] people

```
     187
   ×  32
     374
     561
    5984
```

D. Every rug has 28 pins in it. There are 896 pins in rugs. How many rugs are there?

[32] rugs

```
         32
  28 | 896
     - 84
       56
     - 56
        0
```

Part 2 continues on the next page.

E. Mindy has 247 grasshoppers. 192 hop away. How many grasshoppers are left?

[55] grasshoppers

```
     247
   - 192
      55
```

F. Chuck has 1050 puzzle pieces. Every puzzle is made out of 50 pieces. How many puzzles does Chuck have?

[21] puzzles

```
          21
  50 | 1050
     - 100
        50
      - 50
         0
```

3

A.
```
          94
  23 | 2174
     - 207
       104
      - 92
        12
```

B.
```
         52
  17 | 886
     - 85
       36
     - 34
        2
```

C.
```
          46
  27 | 1242
     - 108
       162
     - 162
         0
```

D.
```
        1403
   7 | 9827
     - 7
       28
     - 28
       27
     - 21
        6
```

E.
```
         181
  23 | 4181
     - 23
       188
     - 184
        41
      - 23
        18
```

F.
```
         25
  13 | 328
     - 26
       68
     - 65
        3
```

G.
```
         606
   4 | 2426
     - 24
        26
      - 24
         2
```

H.
```
         509
  15 | 7642
     - 75
       142
     - 135
         7
```

I.
```
        2953
   3 | 8859
     - 6
       28
     - 27
       15
     - 15
        9
      - 9
        0
```

1

```
    4        2        9        9        9        7        3
4 | 16   6 | 12   7 | 63   3 | 27   5 | 45   6 | 42   8 | 24

    9        3        6        3        8        2        6
6 | 54   4 | 12   9 | 54   3 | 9    7 | 56  10 | 20   6 | 36

    7        8        2        5        3        4        5
1 | 7    3 | 24   4 | 8    5 | 25   6 | 18   8 | 32   3 | 15

   10        2        2        5        4        5        7
10 | 100  8 | 16   3 | 6    4 | 20   1 | 4    6 | 30   4 | 28

    7        2        6        4        4        7        8
3 | 21   5 | 10   4 | 24   6 | 24   3 | 12   7 | 49   6 | 48

    6        9        9        8        8
3 | 18   4 | 36   9 | 81   9 | 72   4 | 32
```

2

A.
```
         61
  15 | 915
     - 90
       15
     - 15
        0
```

B.
```
         24
  37 | 889
     - 74
       149
     - 148
         1
```

C.
```
        111
   7 | 780
     - 7
       8
     - 7
      10
     - 7
       3
```

D.
```
        2308
   3 | 6924
     - 6
       9
     - 9
      24
    - 24
       0
```

E.
```
          52
  88 | 4604
     - 440
       204
     - 176
        28
```

F.
```
          90
  24 | 2181
     - 216
        21
```

G.
```
         125
  11 | 1385
     - 11
        28
      - 22
        65
      - 55
        10
```

H.
```
          86
  23 | 2000
     - 184
       160
     - 138
        22
```

Lesson 65 (continued)

3

A Ernest has 30 boxes. 10 are in stack A. The rest are in stack B. How many are in stack B?

30 − 10 = 20

20 boxes

B Jan makes stacks of boxes. She puts 31 in stack A, and 31 in stack B, and 31 in stack C. How many boxes are there?

31 + 31 + 31 = 93 or 31 × 3 = 93

93 boxes

C Bill can stack 59 boxes in an hour. If he stacks 295 boxes, how many hours did he work?

59⟌295 = 5; 295 − 295 = 0

5 hours

D Martin has 64 boxes in each stack. 12 are in stack A. The rest are in stack B. How many are in stack B?

64 − 12 = 52

52 boxes

E There are 3 boxes in each stack. There are 102 stacks. How many boxes are there?

102 × 3 = 306

306 boxes

F Katie can stack 45 boxes in an hour. If she stacks boxes for 5 hours, how many boxes can she stack?

45 × 5 = 225

225 boxes

G Ana needs 56 boxes. She already has 12. How many boxes does she still need to get?

56 − 12 = 44

44 boxes

Transition Lesson 6

BONUS

1

A 2⟌6 = 3 B 5⟌10 = 2 C 2⟌12 = 6

2

A 5⟌15 B 5⟌10 C 5⟌20 D 5⟌5

3

A 4⟌24 = 6; 6 / 4⟌24; 4 / 6⟌24; 6 / 4⟌24; 4 / 6⟌24

B 3⟌15 = 5; 5 / 3⟌15; 3 / 5⟌15

C 4⟌12 = 3; 3 / 4⟌12; 4 / 3⟌12

4

A 5⟌20 = 4 B 2⟌6 = 3 C 2⟌12 = 6 D 5⟌15 = 3

5

A 5⟌20 = 4 B 5⟌10 = 2 C 5⟌25 = 5

Transition Lesson 27

BONUS

1

A 2⟌6 = 3 B 5⟌10 = 2 C 2⟌12 = 6

2

A 4⟌24 = 6; 6 / 4⟌24; 4 / 6⟌24; 6 / 4⟌24; 4 / 6⟌24

B 3⟌15 = 5; 5 / 3⟌15; 3 / 5⟌15

C 4⟌12 = 3; 3 / 4⟌12; 4 / 3⟌12

3

A 5⟌638 B 5⟌4627 C 5⟌532 D 9⟌467

4

A 5⟌21; 4 / 5⟌20 B 5⟌12; 2 / 5⟌10 C 5⟌16; 3 / 5⟌15

D 5⟌9; 1 / 5⟌5 E 5⟌19; 3 / 5⟌15

5

A 9⟌21; 2 / 9⟌18 B 9⟌12; 1 / 9⟌9 C 9⟌39; 4 / 9⟌36

D 9⟌32; 3 / 9⟌27 E 9⟌11; 1 / 9⟌9

Mastery Test Review—Lesson 20

1

A 5⟌18 = 3; −15; 3
B 9⟌19 = 2; −18; 1
C 5⟌17 = 3; −15; 2
D 9⟌6 = 0; −0; 6
E 9⟌38 = 4; −36; 2
F 5⟌6 = 1; −5; 1

2

A 9⟌22 = 2; −18; 4
B 9⟌33 = 3; −27; 6
C 9⟌44 = 4; −36; 8

3

A 9⟌38 = 4; −36; 2
B 5⟌14 = 2; −10; 4
C 5⟌23 = 4; −20; 3
D 9⟌17 = 1; −9; 8
E 9⟌29 = 3; −27; 2

4

A 9⟌52 = 5; −45; 7
B 9⟌23 = 2; −18; 5
C 9⟌40 = 4; −36; 4

Mastery Test Review—Lesson 27

1

A A dog chewed 6 bones every day. It chewed bones for 90 days. How many bones did the dog chew?

B George cleaned 9 erasers each week. He cleaned erasers for 45 weeks. How many erasers did George clean?

C Marcy washed 3 cars every hour. She washed cars for 60 hours. How many cars did Marcy wash?

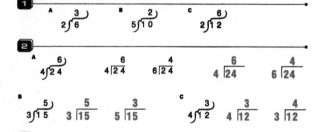

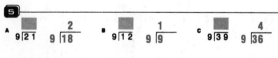

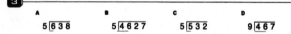

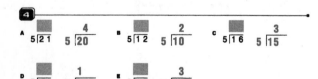

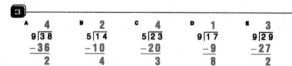

Mastery Test Review Lesson 27 (continued)

2

A. Shu fixed 4 tires every day. He fixed tires for 32 days. How many tires did Shu fix?

B. Mr. Silbert wrote 2 stories every week. He wrote 12 stories. How many weeks did he write stories?

C. Carmen painted 5 cars every week. She painted 10 cars. How many weeks did Carmen paint?

Mastery Test Review—Lesson 30

1

- A. 5)2846 = 569; −25, 34, −30, 46, −45, 1
- B. 9)4730 = 525; −45, 23, −18, 50, −45, 5
- C. 3)7305 = 2435; −6, 13, −12, 10, −9, 15, −15, 0
- D. 9)3790 = 421; −36, 19, −18, 10, −9, 1
- E. 9)1231 = 136; −9, 33, −27, 61, −54, 7
- F. 5)2173 = 434; −20, 17, −15, 23, −20, 3

Mastery Test Review Lesson 30 (continued)

2

- A. 9)2081 = 231; −18, 28, −27, 11, −9, 2
- B. 8)6937 = 867; −64, 53, −48, 57, −56, 1
- C. 8)7904 = 988; −72, 70, −64, 64, −64, 0
- D. 5)2596 = 519; −25, 09, −5, 46, −45, 1
- E. 8)4470 = 558; −40, 47, −40, 70, −64, 6

Mastery Test Review—Lesson 35

1

- A. 9)3624 = 402; −36, 024, −18, 6
- B. 8)4065 = 508; −40, 065, −64, 1
- C. 5)4530 = 906; −45, 030, −30, 0
- D. 3)1210 = 403; −12, 010, −9, 1
- E. 9)1838 = 204; −18, 038, −36, 2

2

- A. 5)203 = 40; −20, 03
- B. 7)356 = 50; −35, 06
- C. 7)74 = 10; −7, 04
- D. 2)161 = 80; −16, 01
- E. 2)81 = 40; −8, 01
- F. 7)213 = 30; −21, 03

Mastery Test Review—Lesson 39

1

A. 146 B. 391 C. 605 D. 992

2

- A. 798 rounds to 80 tens.
- B. 474 rounds to 47 tens.
- C. 203 rounds to 20 tens.
- D. 645 rounds to 65 tens.
- E. 155 rounds to 16 tens.
- F. 297 rounds to 30 tens.
- G. 579 rounds to 58 tens.
- H. 512 rounds to 51 tens.

3

- A. 265 rounds to 27 tens.
- B. 304 rounds to 30 tens.
- C. 737 rounds to 74 tens.
- D. 972 rounds to 97 tens.
- E. 180 rounds to 18 tens.
- F. 846 rounds to 85 tens.
- G. 591 rounds to 59 tens.
- H. 459 rounds to 46 tens.

Mastery Test Review—Lesson 47

1

- A. 6)19 = 3; 6)13 = 2; 58)1872 = 32; −174, 132, −116, 16
- B. 7)10 = 1; 7)26 = 3; 72)984 = 13; −72, 264, −216, 48
- C. 5)9 = 1; 5)47 = 9; 48)948 = 19; −48, 468, −432, 36
- D. 6)26 = 4; 6)8 = 1; 62)2561 = 41; −248, 81, −62, 19

Mastery Test Review Lesson 47 (continued)

2

- A. 6)8 = 1; 6)18 = 3; 57)752 = 13; −57, 182, −171, 11
- B. 2)9 = 4; 2)7 = 3; 22)946 = 43; −88, 66, −66, 0
- C. 4)25 = 6; 4)19 = 4; 39)2528 = 64; −234, 188, −156, 32
- D. 7)17 = 2; 7)33 = 4; 70)1726 = 24; −140, 326, −280, 46

Mastery Test Review—Lesson 50

1

- A. 2)8; 2)13; 22)794 = 36; −66, 134, −132, 2
- B. 4)32; 4)9; 35)3243 = 92; −315, 93, −70, 23
- C. 6)23; 6)9; 56)2327 = 41; −224, 87, −56, 31
- D. 5)25; 5)44; 51)2481 = 48; −204, 441, −408, 33
- E. 7)38; 7)23; 72)3829 = 53; −360, 229, −216, 13

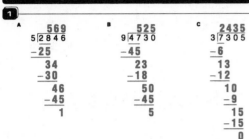

Mastery Test Review Lesson 50 (continued)

2

A Boxes: 5|40, 5|30

```
     75
53|4014
  -371
    304
   -265
     39
```

B Boxes: 7|53, 7|13

```
     81
65|5328
  -520
    128
    -65
     63
```

C Boxes: 2|6, 2|13

```
     25
24|607
  -48
   127
  -120
     7
```

D Boxes: 6|20, 6|12

```
     32
62|1984
  -186
    124
   -124
      0
```

Mastery Test Review—Lesson 54

1

A Boxes: 7|15, 7|33

```
      204
74|15,125
  -148
     325
    -296
      29
```

B Boxes: 6|38, 6|58

```
      609
62|37,781
  -372
     581
    -558
      23
```

C Boxes: 7|56, 7|48

```
      807
69|55,6
  -552
     484
    -483
       1
```

D Boxes: 2|7, 2|11

```
     305
22|6714
  -66
    114
   -110
      4
```


Mastery Test Review Lesson 54 (continued)

2

A Boxes: 6|45

```
     70
64|4496
  -448
     16
```

B Boxes: 7|36

```
     50
71|3598
  -355
     48
```

C Boxes: 4|10, 4|13

```
     22
43|987
  -86
   127
   -86
    41
```

D Boxes: 5|5, 5|22

```
     104
52|5415
  -52
    215
   -208
      7
```

Mastery Test Review—Lesson 60

1

A There are 6 kids on each team. There are 48 kids. How many teams are there?

Answer: **8** teams (+ − × ÷, ÷ circled)

Work: 48 ÷ 6 = 8

B Sherry has some leaves. 8 of them are red and 9 of them are yellow. How many leaves does Sherry have?

Answer: **17** leaves (+ − × ÷, + circled)

Work: 8 + 9 = 17

C The parade has 17 bands. 6 of the bands are on floats and the rest of them are marching. How many bands are marching?

Answer: **11** bands (+ − × ÷, − circled)

Work: 17 − 6 = 11

Part 1 continues on the next page.

Mastery Test Review Lesson 60 (continued)

D Pete did 35 problems on each page. He did 11 pages of problems. How many problems did he do?

Answer: **385** problems (+ − × ÷, × circled)

Work: 35 × 11 = 35 + 350 → 385

E Betty typed 7 pages an hour. She typed 56 pages. How many hours did she type?

Answer: **8** hours (+ − × ÷, ÷ circled)

Work: 56 ÷ 7 = 8

2

A A woman has 126 plants in her garden. Her daughter has 96. How many more plants does the woman have than her daughter?

Answer: **30** plants (+ − × ÷, − circled)

Work: 126 − 96 = 30

B Vic drinks water 6 times every day. After 42 days go by, how many times will Vic have drunk water?

Answer: **252** times (+ − × ÷, × circled)

Work: 42 × 6 = 252

C There are 4 cans in each box. There are 24 cans in boxes. How many boxes are there?

Answer: **6** boxes (+ − × ÷, ÷ circled)

Work: 24 ÷ 4 = 6

D The doctor saw 32 patients on Monday, 20 patients on Tuesday, 4 patients on Wednesday, and 7 patients on Thursday. How many patients did she see?

Answer: **63** patients (+ − × ÷, + circled)

Work: 32 + 20 + 4 + 7 = 63

Part 2 continues on the next page.

Mastery Test Review Lesson 60 (continued)

E Every bag Mike carried weighed 18 kilograms. If he carried 9 bags, how many kilograms did he carry?

Answer: **162** kilograms (+ − × ÷, × circled)

Work: 18 × 9 = 162

F Millie went to the bank 9 times every month. In all, she went to the bank 126 times. How many months did Millie go to the bank?

Answer: **14** months (+ − × ÷, ÷ circled)

```
    14
9|126
 -9
  36
 -36
   0
```

Lesson 3

Name ______________________

9)45 = 5	5)15 = 3	9)9 = 1	9)27 = 3	5)10 = 2	5)25 = 5	9)36 = 4
9)18 = 2	5)20 = 4	5)5 = 1	9)45 = 5	9)18 = 2	5)15 = 3	5)5 = 1
5)20 = 4	9)9 = 1	9)27 = 3	9)36 = 4	5)25 = 5	5)10 = 2	9)36 = 4
9)9 = 1	5)20 = 4	9)18 = 2	9)45 = 5	5)5 = 1	5)15 = 3	5)25 = 5
9)27 = 3	9)45 = 5	5)10 = 2	9)27 = 3	9)36 = 4	9)9 = 1	5)15 = 3

Lesson 6

Name ______________________

1)10 = 10	1)3 = 3	9)45 = 5	9)18 = 2	1)4 = 4	1)8 = 8	5)5 = 1
9)9 = 1	1)1 = 1	1)7 = 7	1)2 = 2	9)27 = 3	1)6 = 6	5)25 = 5
1)5 = 5	1)9 = 9	1)5 = 5	5)10 = 2	9)18 = 2	1)10 = 10	1)6 = 6
9)9 = 1	5)20 = 4	1)1 = 1	1)4 = 4	5)15 = 3	9)36 = 4	1)2 = 2
1)9 = 9	5)15 = 3	1)3 = 3	5)10 = 2	1)8 = 8	5)5 = 1	1)7 = 7

Lesson 8

Name ______________________

1)10 = 10	9)45 = 5	9)9 = 1	5)10 = 2	1)8 = 8	1)3 = 3	9)36 = 4
9)27 = 3	5)20 = 4	1)4 = 4	5)5 = 1	1)9 = 9	9)27 = 3	1)1 = 1
1)6 = 6	5)10 = 2	9)18 = 2	1)2 = 2	1)5 = 5	5)25 = 5	1)7 = 7
1)1 = 1	9)9 = 1	1)3 = 3	1)8 = 8	5)15 = 3	1)4 = 4	1)7 = 7
5)20 = 4	1)6 = 6	9)36 = 4	1)5 = 5	1)10 = 10	9)18 = 2	1)9 = 9

Lesson 10

Name ______________________

5)5 = 1	9)36 = 4	1)2 = 2	1)7 = 7	5)15 = 3	1)5 = 5	1)9 = 9
9)18 = 2	1)1 = 1	1)4 = 4	5)25 = 5	1)3 = 3	1)10 = 10	5)20 = 4
9)45 = 5	1)6 = 6	1)8 = 8	5)15 = 3	1)3 = 3	9)18 = 2	1)6 = 6
1)1 = 1	9)27 = 3	5)10 = 2	1)4 = 4	1)9 = 9	9)36 = 4	1)7 = 7
5)25 = 5	1)8 = 8	1)10 = 10	9)9 = 1	5)5 = 1	1)2 = 2	1)5 = 5

Lesson 13

Name ______________________

5 9⟌45	1 1⟌1	5 5⟌25	7 1⟌7	3 1⟌3	4 9⟌36	1 9⟌9
2 5⟌10	4 5⟌20	5 1⟌5	8 1⟌8	10 1⟌10	1 5⟌5	2 9⟌18
3 5⟌15	3 9⟌27	2 1⟌2	7 1⟌7	4 5⟌20	4 9⟌36	9 1⟌9
2 5⟌10	5 5⟌25	5 9⟌45	3 1⟌3	8 1⟌8	2 9⟌18	1 5⟌5
4 1⟌4	9 1⟌9	1 9⟌9	3 5⟌15	6 1⟌6	3 9⟌27	5 1⟌5

Lesson 16

Name ______________________

5 3⟌15	2 3⟌6	2 1⟌2	4 9⟌36	1 5⟌5	4 5⟌20	4 3⟌12
1 3⟌3	1 9⟌9	5 9⟌45	6 1⟌6	3 3⟌9	4 3⟌12	2 5⟌10
2 9⟌18	1 1⟌1	1 3⟌3	5 3⟌15	5 9⟌45	5 5⟌25	2 3⟌6
3 3⟌9	4 1⟌4	2 9⟌18	3 9⟌27	3 5⟌15	2 3⟌6	5 3⟌15
3 9⟌27	10 1⟌10	4 3⟌12	4 9⟌36	3 5⟌15	1 3⟌3	3 3⟌9

Lesson 18

Name ______________________

10 5⟌50	7 5⟌35	2 3⟌6	7 1⟌7	2 9⟌18	5 5⟌25	8 5⟌40
4 3⟌12	6 5⟌30	9 5⟌45	4 5⟌20	4 1⟌4	4 3⟌12	9 5⟌45
1 3⟌3	9 1⟌9	6 5⟌30	10 5⟌50	2 5⟌10	1 9⟌9	8 5⟌40
3 3⟌9	7 5⟌35	1 5⟌5	4 9⟌36	6 5⟌30	5 3⟌15	10 5⟌50
3 3⟌9	1 3⟌3	8 5⟌40	2 3⟌6	9 5⟌45	5 3⟌15	7 5⟌35

Lesson 20

Name ______________________

6 5⟌30	9 5⟌45	1 9⟌9	5 1⟌5	2 3⟌6	7 5⟌35	10 5⟌50
5 9⟌45	1 3⟌3	8 5⟌40	1 5⟌5	2 9⟌18	7 5⟌35	5 3⟌15
8 1⟌8	9 5⟌45	3 9⟌27	10 5⟌50	10 1⟌10	3 3⟌9	6 5⟌30
8 5⟌40	4 9⟌36	4 5⟌20	4 3⟌12	7 5⟌35	3 5⟌15	5 3⟌15
2 5⟌10	9 5⟌45	6 5⟌30	4 3⟌12	5 5⟌25	8 5⟌40	10 5⟌50

Lesson 23 Name ______________________

8)80 = 10	5)45 = 9	8)56 = 7	8)40 = 5	9)36 = 4	5)15 = 3	3)3 = 1
8)48 = 6	8)72 = 9	9)18 = 2	3)6 = 2	8)64 = 8	5)50 = 10	5)35 = 7
3)12 = 4	9)45 = 5	5)25 = 5	8)48 = 6	5)30 = 6	8)72 = 9	8)56 = 7
5)20 = 4	8)40 = 5	8)64 = 8	8)80 = 10	5)40 = 8	3)15 = 5	8)56 = 7
8)40 = 5	3)9 = 3	8)72 = 9	9)27 = 3	8)64 = 8	8)80 = 10	8)48 = 6

Lesson 26 Name ______________________

5)5 = 1	8)80 = 10	5)45 = 9	3)15 = 5	8)40 = 5	8)64 = 8	5)30 = 6
9)18 = 2	8)48 = 6	5)10 = 2	5)35 = 7	8)72 = 9	8)56 = 7	5)40 = 8
3)9 = 3	1)2 = 2	8)48 = 6	5)15 = 3	9)36 = 4	1)6 = 6	8)64 = 8
8)80 = 10	3)12 = 4	1)3 = 3	9)45 = 5	5)20 = 4	8)40 = 5	8)72 = 9
3)6 = 2	9)9 = 1	8)56 = 7	5)50 = 10	5)25 = 5	9)27 = 3	3)3 = 1

Lesson 28 Name ______________________

2)12 = 6	2)4 = 2	8)56 = 7	5)35 = 7	2)10 = 5	2)2 = 1	3)15 = 5
2)6 = 3	8)40 = 5	2)8 = 4	8)72 = 9	2)12 = 6	1)1 = 1	2)10 = 5
2)4 = 2	8)64 = 8	5)50 = 10	2)6 = 3	8)80 = 10	3)9 = 3	2)8 = 4
2)2 = 1	3)3 = 1	2)4 = 2	2)10 = 5	3)6 = 2	5)30 = 6	2)8 = 4
5)40 = 8	2)6 = 3	3)12 = 4	2)2 = 1	8)48 = 6	2)12 = 6	5)45 = 9

Lesson 30 Name ______________________

2)20 = 10	2)14 = 7	1)9 = 9	2)2 = 1	2)16 = 8	8)48 = 6	2)10 = 5
2)18 = 9	9)27 = 3	2)8 = 4	2)4 = 2	8)64 = 8	9)45 = 5	2)6 = 3
2)16 = 8	2)20 = 10	8)40 = 5	9)9 = 1	2)14 = 7	8)80 = 10	2)18 = 9
2)12 = 6	8)72 = 9	9)36 = 4	2)20 = 10	2)6 = 3	2)10 = 5	8)56 = 7
2)8 = 4	2)12 = 6	2)16 = 8	2)2 = 1	2)14 = 7	2)4 = 2	2)18 = 9

Lesson 32

Name ____________________

7⟌14 = 2	2⟌20 = 10	2⟌8 = 4	8⟌72 = 9	5⟌40 = 8	3⟌6 = 2	7⟌7 = 1
2⟌16 = 8	2⟌2 = 1	8⟌80 = 10	5⟌30 = 6	2⟌14 = 7	7⟌7 = 1	2⟌6 = 3
8⟌64 = 8	5⟌35 = 7	3⟌12 = 4	2⟌18 = 9	7⟌14 = 2	8⟌56 = 7	3⟌15 = 5
2⟌20 = 10	2⟌12 = 6	8⟌48 = 6	3⟌9 = 3	2⟌10 = 5	2⟌16 = 8	7⟌7 = 1
8⟌40 = 5	5⟌45 = 9	2⟌14 = 7	2⟌18 = 9	7⟌14 = 2	2⟌4 = 2	9⟌45 = 5

Lesson 34

Name ____________________

7⟌35 = 5	7⟌21 = 3	2⟌18 = 9	2⟌4 = 2	8⟌64 = 8	9⟌27 = 3	7⟌28 = 4
7⟌7 = 1	2⟌16 = 8	2⟌6 = 3	8⟌72 = 9	8⟌40 = 5	9⟌36 = 4	2⟌14 = 7
7⟌21 = 3	7⟌35 = 5	8⟌56 = 7	2⟌8 = 4	7⟌14 = 2	2⟌18 = 9	2⟌12 = 6
8⟌80 = 10	8⟌48 = 6	2⟌14 = 7	7⟌35 = 5	7⟌28 = 4	2⟌16 = 8	2⟌2 = 1
7⟌21 = 3	2⟌20 = 10	7⟌14 = 2	2⟌20 = 10	2⟌10 = 5	7⟌28 = 4	9⟌9 = 1

Lesson 36

Name ____________________

4⟌40 = 10	4⟌24 = 6	7⟌21 = 3	8⟌72 = 9	3⟌9 = 3	4⟌32 = 8	4⟌24 = 6
7⟌35 = 5	8⟌48 = 6	4⟌28 = 7	7⟌35 = 5	4⟌32 = 8	9⟌18 = 2	4⟌28 = 7
4⟌40 = 10	7⟌28 = 4	7⟌21 = 3	5⟌25 = 5	9⟌45 = 5	4⟌36 = 9	7⟌14 = 2
5⟌50 = 10	4⟌32 = 8	7⟌28 = 4	5⟌15 = 3	5⟌35 = 7	4⟌28 = 7	7⟌14 = 2
4⟌40 = 10	5⟌20 = 4	3⟌15 = 5	4⟌24 = 6	5⟌45 = 9	4⟌36 = 9	5⟌45 = 9

Lesson 38

Name ____________________

10⟌80 = 8	10⟌30 = 3	4⟌36 = 9	7⟌28 = 4	10⟌100 = 10	10⟌40 = 4	4⟌32 = 8
2⟌12 = 6	4⟌28 = 7	10⟌50 = 5	10⟌90 = 9	10⟌10 = 1	4⟌24 = 6	7⟌7 = 1
10⟌20 = 2	10⟌70 = 7	4⟌40 = 10	10⟌60 = 6	7⟌35 = 5	2⟌20 = 10	10⟌90 = 9
2⟌14 = 7	4⟌24 = 6	10⟌60 = 6	10⟌10 = 1	7⟌21 = 3	2⟌16 = 8	10⟌100 = 10
10⟌40 = 4	2⟌18 = 9	10⟌20 = 2	10⟌80 = 8	7⟌14 = 2	10⟌70 = 7	10⟌30 = 3

Lesson 40

Name ______________________

5 6)30	1 6)6	7 10)70	7 4)28	3 6)18	10 10)100	4 6)24
1 10)10	2 6)12	8 4)32	4 6)24	8 10)80	2 10)20	2 6)12
5 6)30	3 10)30	9 4)36	1 6)6	3 6)18	9 10)90	4 10)40
10 4)40	4 6)24	1 6)6	5 10)50	2 6)12	5 6)30	6 10)60
9 10)90	3 6)18	9 10)90	5 10)50	1 6)6	8 10)80	7 10)70

Lesson 42

Name ______________________

5 6)30	10 10)100	6 4)24	3 6)18	3 2)6	2 6)12	6 5)30
5 8)40	2 6)12	9 10)90	4 6)24	9 4)36	8 2)16	2 5)10
1 9)9	3 6)18	7 4)28	3 7)21	8 5)40	2 6)12	2 3)6
5 2)10	4 6)24	10 2)20	7 8)56	4 7)28	1 6)6	8 10)80
4 9)36	4 3)12	5 7)35	8 4)32	8 8)64	10 8)80	5 6)30

Lesson 44

Name ______________________

10 7)70	7 7)49	1 6)6	4 10)40	9 7)63	6 7)42	6 10)60
1 5)5	8 7)56	8 10)80	9 7)63	4 2)8	9 5)45	8 7)56
9 10)90	7 7)49	5 5)25	6 8)48	9 7)63	6 7)42	9 8)72
4 6)24	10 10)100	7 7)49	7 2)14	9 7)63	9 2)18	2 6)12
8 7)56	3 6)18	6 7)42	2 7)14	5 6)30	10 7)70	5 9)45

Lesson 46

Name ______________________

5 4)20	3 10)30	1 7)7	3 4)12	1 4)4	2 10)20	6 2)12
4 5)20	2 4)8	4 4)16	1 10)10	2 2)4	2 4)8	5 4)20
1 15)15	2 9)18	3 3)9	1 4)4	4 4)16	10 5)50	10 4)40
2 4)8	5 4)20	3 9)27	1 2)2	5 3)15	10 7)70	3 4)12
7 5)35	6 7)42	2 4)8	8 7)56	4 4)16	7 7)49	9 7)63

Lesson 48 Name ____________________

4)20 = 5	6)54 = 9	6)30 = 5	10)80 = 8	7)28 = 4	6)48 = 8	6)36 = 6
4)16 = 4	7)63 = 9	6)42 = 7	10)90 = 9	7)35 = 5	6)6 = 1	4)4 = 1
6)48 = 8	10)100 = 10	6)18 = 3	6)42 = 7	7)70 = 10	7)42 = 6	6)36 = 6
6)60 = 10	4)24 = 6	10)70 = 7	7)49 = 7	4)12 = 3	7)56 = 8	6)24 = 4
4)32 = 8	4)24 = 6	6)12 = 2	4)36 = 9	6)54 = 9	6)60 = 10	9)36 = 4

Lesson 50 Name ____________________

6)54 = 9	4)12 = 3	7)42 = 6	6)24 = 4	7)14 = 2	3)12 = 4	8)64 = 8
6)48 = 8	7)70 = 10	6)30 = 5	7)21 = 3	4)16 = 4	6)60 = 10	7)49 = 7
6)12 = 2	2)20 = 10	6)42 = 7	8)40 = 5	7)56 = 8	6)36 = 6	6)18 = 3
10)50 = 5	4)4 = 1	7)63 = 9	6)42 = 7	6)6 = 1	5)40 = 8	8)56 = 7
4)8 = 2	6)36 = 6	6)54 = 9	4)20 = 5	6)48 = 8	6)60 = 10	5)30 = 6

Lesson 52 Name ____________________

9)90 = 10	6)54 = 9	4)12 = 3	7)42 = 6	6)12 = 2	9)81 = 9	4)16 = 4
9)63 = 7	6)18 = 3	7)49 = 7	9)72 = 8	6)42 = 7	9)54 = 6	7)63 = 9
4)4 = 1	6)24 = 4	9)81 = 9	4)20 = 5	9)72 = 8	7)70 = 10	6)30 = 5
9)63 = 7	9)90 = 10	6)48 = 8	7)56 = 8	9)54 = 6	9)81 = 9	4)8 = 2
6)36 = 6	9)90 = 10	9)72 = 8	6)60 = 10	9)54 = 6	4)36 = 9	9)63 = 7

Lesson 54 Name ____________________

3)18 = 6	3)24 = 8	9)81 = 9	10)60 = 6	7)28 = 4	3)21 = 7	9)54 = 6
2)18 = 9	3)27 = 9	6)6 = 1	3)18 = 6	9)45 = 5	3)6 = 2	3)21 = 7
5)45 = 9	8)48 = 6	3)24 = 8	9)63 = 7	3)27 = 9	8)72 = 9	8)80 = 10
9)72 = 8	3)21 = 7	3)15 = 5	4)24 = 6	3)24 = 8	4)32 = 8	9)90 = 10
3)27 = 9	7)28 = 4	2)10 = 5	7)7 = 1	2)8 = 4	7)35 = 5	3)18 = 6

Lesson 56

Name ______________________

$\overset{4}{8\overline{)32}}$	$\overset{10}{3\overline{)30}}$	$\overset{9}{9\overline{)81}}$	$\overset{10}{6\overline{)60}}$	$\overset{2}{4\overline{)8}}$	$\overset{3}{8\overline{)24}}$	$\overset{9}{3\overline{)27}}$
$\overset{8}{7\overline{)56}}$	$\overset{3}{7\overline{)21}}$	$\overset{2}{8\overline{)16}}$	$\overset{8}{3\overline{)24}}$	$\overset{8}{9\overline{)72}}$	$\overset{1}{8\overline{)8}}$	$\overset{7}{3\overline{)21}}$
$\overset{5}{4\overline{)20}}$	$\overset{6}{9\overline{)54}}$	$\overset{6}{3\overline{)18}}$	$\overset{1}{8\overline{)8}}$	$\overset{4}{8\overline{)32}}$	$\overset{7}{7\overline{)49}}$	$\overset{10}{10\overline{)100}}$
$\overset{8}{2\overline{)16}}$	$\overset{3}{8\overline{)24}}$	$\overset{10}{3\overline{)30}}$	$\overset{10}{9\overline{)90}}$	$\overset{8}{3\overline{)24}}$	$\overset{6}{6\overline{)36}}$	$\overset{1}{4\overline{)4}}$
$\overset{6}{3\overline{)18}}$	$\overset{9}{3\overline{)27}}$	$\overset{9}{6\overline{)54}}$	$\overset{3}{7\overline{)21}}$	$\overset{7}{3\overline{)21}}$	$\overset{2}{8\overline{)16}}$	$\overset{7}{9\overline{)63}}$

Lesson 58

Name ______________________

$\overset{3}{8\overline{)24}}$	$\overset{7}{3\overline{)21}}$	$\overset{8}{6\overline{)48}}$	$\overset{2}{8\overline{)16}}$	$\overset{8}{3\overline{)24}}$	$\overset{3}{4\overline{)12}}$	$\overset{3}{6\overline{)18}}$
$\overset{4}{10\overline{)40}}$	$\overset{4}{2\overline{)8}}$	$\overset{10}{3\overline{)30}}$	$\overset{4}{8\overline{)32}}$	$\overset{6}{3\overline{)18}}$	$\overset{7}{2\overline{)14}}$	$\overset{2}{6\overline{)12}}$
$\overset{1}{8\overline{)8}}$	$\overset{9}{3\overline{)27}}$	$\overset{10}{7\overline{)70}}$	$\overset{6}{7\overline{)42}}$	$\overset{4}{6\overline{)24}}$	$\overset{9}{7\overline{)63}}$	$\overset{1}{8\overline{)8}}$
$\overset{4}{8\overline{)32}}$	$\overset{10}{3\overline{)30}}$	$\overset{7}{6\overline{)42}}$	$\overset{4}{4\overline{)16}}$	$\overset{5}{6\overline{)30}}$	$\overset{5}{9\overline{)45}}$	$\overset{2}{8\overline{)16}}$
$\overset{1}{3\overline{)3}}$	$\overset{10}{5\overline{)50}}$	$\overset{8}{8\overline{)64}}$	$\overset{3}{8\overline{)24}}$	$\overset{7}{5\overline{)35}}$	$\overset{3}{3\overline{)9}}$	$\overset{7}{8\overline{)56}}$

Lesson 60

Name ______________________

$\overset{5}{5\overline{)25}}$	$\overset{6}{2\overline{)12}}$	$\overset{4}{8\overline{)32}}$	$\overset{8}{3\overline{)24}}$	$\overset{8}{9\overline{)72}}$	$\overset{5}{4\overline{)20}}$	$\overset{2}{2\overline{)4}}$
$\overset{7}{3\overline{)21}}$	$\overset{1}{8\overline{)8}}$	$\overset{6}{6\overline{)36}}$	$\overset{7}{7\overline{)49}}$	$\overset{1}{10\overline{)10}}$	$\overset{2}{9\overline{)18}}$	$\overset{10}{3\overline{)30}}$
$\overset{10}{9\overline{)90}}$	$\overset{9}{6\overline{)54}}$	$\overset{1}{4\overline{)4}}$	$\overset{3}{10\overline{)30}}$	$\overset{1}{9\overline{)9}}$	$\overset{2}{8\overline{)16}}$	$\overset{6}{9\overline{)54}}$
$\overset{2}{4\overline{)8}}$	$\overset{1}{6\overline{)6}}$	$\overset{3}{9\overline{)27}}$	$\overset{4}{3\overline{)12}}$	$\overset{3}{8\overline{)24}}$	$\overset{6}{3\overline{)18}}$	$\overset{7}{9\overline{)63}}$
$\overset{10}{6\overline{)60}}$	$\overset{8}{7\overline{)56}}$	$\overset{9}{3\overline{)27}}$	$\overset{9}{9\overline{)81}}$	$\overset{1}{2\overline{)2}}$	$\overset{5}{8\overline{)40}}$	$\overset{6}{5\overline{)30}}$

Lesson 62

Name ______________________

$\overset{3}{8\overline{)24}}$	$\overset{8}{9\overline{)72}}$	$\overset{7}{6\overline{)42}}$	$\overset{3}{4\overline{)12}}$	$\overset{7}{7\overline{)49}}$	$\overset{7}{5\overline{)35}}$	$\overset{2}{3\overline{)6}}$
$\overset{4}{5\overline{)20}}$	$\overset{1}{8\overline{)8}}$	$\overset{8}{3\overline{)24}}$	$\overset{6}{9\overline{)54}}$	$\overset{4}{4\overline{)16}}$	$\overset{2}{5\overline{)10}}$	$\overset{5}{1\overline{)5}}$
$\overset{9}{5\overline{)45}}$	$\overset{10}{3\overline{)30}}$	$\overset{9}{9\overline{)81}}$	$\overset{10}{7\overline{)70}}$	$\overset{6}{7\overline{)42}}$	$\overset{1}{5\overline{)5}}$	$\overset{2}{6\overline{)12}}$
$\overset{4}{8\overline{)32}}$	$\overset{10}{9\overline{)90}}$	$\overset{7}{3\overline{)21}}$	$\overset{9}{7\overline{)63}}$	$\overset{6}{8\overline{)48}}$	$\overset{3}{5\overline{)15}}$	$\overset{2}{8\overline{)16}}$
$\overset{9}{3\overline{)27}}$	$\overset{7}{9\overline{)63}}$	$\overset{8}{6\overline{)48}}$	$\overset{6}{3\overline{)18}}$	$\overset{5}{6\overline{)30}}$	$\overset{9}{8\overline{)72}}$	$\overset{4}{6\overline{)24}}$

Lesson 64

Name ______________________

$\begin{array}{r}10\\3\overline{)30}\end{array}$	$\begin{array}{r}2\\8\overline{)16}\end{array}$	$\begin{array}{r}6\\6\overline{)36}\end{array}$	$\begin{array}{r}1\\4\overline{)4}\end{array}$	$\begin{array}{r}6\\7\overline{)42}\end{array}$	$\begin{array}{r}5\\9\overline{)45}\end{array}$	$\begin{array}{r}5\\3\overline{)15}\end{array}$
$\begin{array}{r}1\\8\overline{)8}\end{array}$	$\begin{array}{r}9\\6\overline{)54}\end{array}$	$\begin{array}{r}2\\4\overline{)8}\end{array}$	$\begin{array}{r}8\\7\overline{)56}\end{array}$	$\begin{array}{r}7\\8\overline{)56}\end{array}$	$\begin{array}{r}10\\8\overline{)80}\end{array}$	$\begin{array}{r}7\\1\overline{)7}\end{array}$
$\begin{array}{r}1\\3\overline{)3}\end{array}$	$\begin{array}{r}5\\4\overline{)20}\end{array}$	$\begin{array}{r}4\\8\overline{)32}\end{array}$	$\begin{array}{r}10\\10\overline{)100}\end{array}$	$\begin{array}{r}9\\4\overline{)36}\end{array}$	$\begin{array}{r}4\\7\overline{)28}\end{array}$	$\begin{array}{r}5\\2\overline{)10}\end{array}$
$\begin{array}{r}6\\4\overline{)24}\end{array}$	$\begin{array}{r}2\\10\overline{)20}\end{array}$	$\begin{array}{r}10\\2\overline{)20}\end{array}$	$\begin{array}{r}3\\3\overline{)9}\end{array}$	$\begin{array}{r}10\\5\overline{)50}\end{array}$	$\begin{array}{r}8\\8\overline{)64}\end{array}$	$\begin{array}{r}7\\4\overline{)28}\end{array}$
$\begin{array}{r}5\\10\overline{)50}\end{array}$	$\begin{array}{r}8\\4\overline{)32}\end{array}$	$\begin{array}{r}1\\7\overline{)7}\end{array}$	$\begin{array}{r}3\\2\overline{)6}\end{array}$	$\begin{array}{r}5\\7\overline{)35}\end{array}$	$\begin{array}{r}3\\8\overline{)24}\end{array}$	$\begin{array}{r}4\\9\overline{)36}\end{array}$

Lesson 2

Name ______________________

$\begin{array}{r}5\\\times 3\\\hline 15\end{array}$	$\begin{array}{r}5\\\times 4\\\hline 20\end{array}$	$\begin{array}{r}9\\\times 1\\\hline 9\end{array}$	$\begin{array}{r}5\\\times 5\\\hline 25\end{array}$	$\begin{array}{r}9\\\times 3\\\hline 27\end{array}$	$\begin{array}{r}9\\\times 5\\\hline 45\end{array}$	$\begin{array}{r}9\\\times 4\\\hline 36\end{array}$

$\begin{array}{r}514\\-403\\\hline 111\end{array}$	$\begin{array}{r}877\\-654\\\hline 223\end{array}$	$\begin{array}{r}225\\-111\\\hline 114\end{array}$	$\begin{array}{r}364\\-363\\\hline 1\end{array}$	$\begin{array}{r}879\\-64\\\hline 815\end{array}$
$\begin{array}{r}323\\+744\\\hline 1067\end{array}$	$\begin{array}{r}312\\-281\\\hline 31\end{array}$	$\begin{array}{r}743\\+731\\\hline 1474\end{array}$	$\begin{array}{r}978\\+322\\\hline 1300\end{array}$	$\begin{array}{r}565\\-221\\\hline 344\end{array}$

A Frank had 14 model cars in his room. He built more model cars. He ended up with 24 model cars. How many model cars did Frank build?

$\begin{array}{r}24\\-14\\\hline 10\end{array}$ model cars

B Chloe liked growing things in her garden. She planted 2 rows of corn, 3 rows of beans, and 2 rows of carrots. How many rows did Chloe have in her garden?

$\begin{array}{r}2\\3\\+2\\\hline 7\end{array}$ rows

C Francine had a large book collection. The books were written by men and women. If 24 books were written by men and 31 books were written by women, how many books did Francine have in all?

$\begin{array}{r}24\\+31\\\hline 55\end{array}$ books

Lesson 4

Name ______________________

$\begin{array}{r}9\\\times 2\\\hline 18\end{array}$	$\begin{array}{r}5\\\times 5\\\hline 25\end{array}$	$\begin{array}{r}9\\\times 5\\\hline 45\end{array}$	$\begin{array}{r}5\\\times 3\\\hline 15\end{array}$	$\begin{array}{r}9\\\times 1\\\hline 9\end{array}$	$\begin{array}{r}5\\\times 4\\\hline 20\end{array}$	$\begin{array}{r}9\\\times 3\\\hline 27\end{array}$

$\begin{array}{r}640\\+721\\\hline 1361\end{array}$	$\begin{array}{r}312\\+974\\\hline 1286\end{array}$	$\begin{array}{r}445\\+221\\\hline 666\end{array}$	$\begin{array}{r}674\\+550\\\hline 1224\end{array}$	$\begin{array}{r}313\\+464\\\hline 777\end{array}$
$\begin{array}{r}441\\-22\\\hline 419\end{array}$	$\begin{array}{r}6745\\-178\\\hline 6567\end{array}$	$\begin{array}{r}945\\-809\\\hline 136\end{array}$	$\begin{array}{r}4170\\-9\\\hline 4161\end{array}$	$\begin{array}{r}200\\-88\\\hline 112\end{array}$

8 × 44 = 352 $\begin{array}{r}44\\\times 8\\\hline 352\end{array}$

6 × 43 = 258 $\begin{array}{r}43\\\times 6\\\hline 258\end{array}$

9 × 81 = 729 $\begin{array}{r}81\\\times 9\\\hline 729\end{array}$

5 × 43 = 315 $\begin{array}{r}63\\\times 5\\\hline 315\end{array}$

7 × 52 = 364 $\begin{array}{r}52\\\times 7\\\hline 364\end{array}$

3 × 20 = 60 $\begin{array}{r}20\\\times 3\\\hline 60\end{array}$

Lesson 6

Name ______________________

$\begin{array}{r}1\\\times 8\\\hline 8\end{array}$	$\begin{array}{r}5\\\times 3\\\hline 15\end{array}$	$\begin{array}{r}9\\\times 2\\\hline 18\end{array}$	$\begin{array}{r}1\\\times 9\\\hline 9\end{array}$	$\begin{array}{r}5\\\times 5\\\hline 25\end{array}$	$\begin{array}{r}9\\\times 4\\\hline 36\end{array}$	$\begin{array}{r}1\\\times 5\\\hline 5\end{array}$

$\begin{array}{r}645\\-564\\\hline 81\end{array}$	$\begin{array}{r}9732\\-2120\\\hline 7612\end{array}$	$\begin{array}{r}5150\\-1547\\\hline 3603\end{array}$	$\begin{array}{r}63\\-48\\\hline 15\end{array}$	$\begin{array}{r}9741\\-2999\\\hline 6742\end{array}$
$\begin{array}{r}625\\+333\\\hline 958\end{array}$	$\begin{array}{r}9647\\-480\\\hline 9167\end{array}$	$\begin{array}{r}5671\\+212\\\hline 5883\end{array}$	$\begin{array}{r}9748\\-2214\\\hline 7534\end{array}$	$\begin{array}{r}3440\\+7984\\\hline 11424\end{array}$

A Joe liked peanut butter and pickle sandwiches. Every day, his mom put 8 peanut butter and pickle sandwiches in his lunchbox. If each sandwich had 4 pickles, how many pickles were there in Joe's lunchbox every day?

$\begin{array}{r}8\\\times 4\\\hline 32\end{array}$ pickles

B In Jennifer's town, each empty soda can could be recycled for 5 cents. If Jennifer recycled 532 cans, how many cents would she have?

$\begin{array}{r}532\\\times 5\\\hline 2660\end{array}$ cents

C Cassie had 54 shirts. 21 of the shirts were purple. How many shirts were not purple?

$\begin{array}{r}54\\-21\\\hline 33\end{array}$ shirts

Lesson 8

Name ____________

A 5) 5 = [1] | 5) 5 = 1 | 1) 5 = 5

B 1) 7 = [7] | 1) 7 = 7 | 7) 7 = 1

C 1) 4 = [4] | 1) 4 = 4 | 4) 4 = 1

D 5) 10 = [2] | 5) 10 = 2 | 2) 10 = 5

5 × 3 = 15	1 × 9 = 9	9 × 5 = 45	5 × 5 = 25	1 × 7 = 7	5 × 4 = 20	9 × 4 = 36

A A box of crackers has 4 packages. If there are 32 crackers in each package, how many crackers are there in a box?

32 × 4 = 128 crackers

B Each airplane seats 148 people. If there are 3 full airplanes, how many people are there?

148 × 3 = 444 people

480 + 671 = 1151	228 − 134 = 94	672 − 595 = 77	780 − 364 = 416	915 + 451 = 1366
6714 + 2347 = 9061	3719 + 2555 = 6274	1974 − 1972 = 2	9764 − 4646 = 5118	3210 + 3714 = 6924

Lesson 10

Name ____________

1 × 8 = 8	9 × 2 = 18	1 × 5 = 5	5 × 3 = 15	9 × 5 = 45	5 × 5 = 25	1 × 3 = 3

A Francis had 40 crayons. 18 crayons were red. How many crayons were not red?

40 − 18 = 22 crayons

B Each student in the class brought 5 boxes of tissues at the beginning of the school year. If there were 29 students in the class, how many boxes of tissues were there altogether?

29 × 5 = 145 boxes

C Mr. Jackson drove 248 miles. He stopped to eat lunch. He got back into his car and drove 572 more miles. Then he stopped for dinner. He drove 280 more miles before stopping for the night. How many miles did Mr. Jackson drive in all?

248 + 572 + 280 = 1100 miles

A 5) 25 = [5] | 5) 25 = 5 | 5) 25 = 5

B 5) 10 = [2] | 5) 10 = 2 | 2) 10 = 5

C 1) 7 = [7] | 1) 7 = 7 | 7) 7 = 1

D 1) 5 = [5] | 1) 5 = 5 | 5) 5 = 1

Lesson 12

Name ____________

5) 15 = 3	9) 45 = 5	5) 25 = 5	9) 36 = 4	1) 10 = 10
1) 2 = 2	9) 18 = 2	1) 7 = 7	5) 20 = 4	9) 9 = 1

9 × 5 = 45	1 × 1 = 1	5 × 4 = 20	9 × 2 = 18	1 × 9 = 9	9 × 3 = 27	5 × 5 = 25

452 − 211 = 241	515 + 515 = 1030	978 − 564 = 414	244 + 754 = 998	349 − 287 = 62
2371 + 3789 = 6160	6471 − 5998 = 473	3574 + 1597 = 5171	6397 − 4571 = 1826	1257 + 1348 = 2605

A Keith answered 50 questions on his test correctly. If there were 57 questions on the test, how many did Keith answer incorrectly?

57 − 50 = 7 questions

B Lola, the dog, loved to go for walks. On Monday she walked for 3 miles. She chased 8 cats. She walked 2 more miles and sniffed 3 trees. How many miles did Lola walk on Monday?

3 + 2 = 5 miles

Lesson 14

Name ____________

9) 18 = 2	5) 5 = 1	9) 9 = 1	5) 10 = 2	9) 36 = 4
9) 27 = 3	1) 4 = 4	1) 8 = 8	1) 6 = 6	5) 25 = 5

5 × 5 = 25	9 × 4 = 36	1 × 4 = 4	5 × 2 = 10	9 × 3 = 27	5 × 3 = 15	1 × 5 = 5

453 × 9 = 4077	874 × 5 = 4370	256 × 2 = 512	354 × 8 = 2832	541 × 5 = 2705	424 × 8 = 3392

A DeShaun has 45 books on each shelf in his room. If he has 9 shelves, how many books does DeShaun have in all?

45 × 9 = 405 books

B Each table tennis ball weighed 2 ounces. If there were 945 table tennis balls, how much would they weigh altogether?

945 × 2 = 1890 ounces

C A pitcher threw a baseball 90 miles per hour. The next pitch was 102 miles per hour. How much faster was the second pitch than the first?

102 − 90 = 12 miles per hour

Lesson 16

Name ____________________

5)20 = 4	9)36 = 4	1)4 = 4	9)45 = 5	9)18 = 2
9)9 = 1	9)27 = 3	5)10 = 2	5)5 = 1	5)15 = 3

24 × 7 = 168	16 × 5 = 80	43 × 9 = 387	17 × 5 = 85	32 × 6 = 192	48 × 3 = 144

764 − 541 = 223	3814 + 2121 = 5935	7348 + 1247 = 8595	9742 + 3201 = 12,943	6440 − 3894 = 2546
3174 − 2451 = 723	874 + 22 = 896	9466 − 2510 = 6956	741 − 603 = 138	5341 + 9741 = 15,082

A One small number in a times problem is 5. The other small number is 3. What is the third number?
5 × 3 = 15

B Oliver went to the store to buy some things. He put 78 items into his cart. 9 of the items were not food. How many of the items were food?
78 − 9 = 69 items

C One small number in a times problem is 9. The other small number is 4. What is the third number?
9 × 4 = 36

Lesson 18

Name ____________________

3)12 = 4	3)15 = 5	3)9 = 3	1)10 = 10	1)7 = 7
5)5 = 1	5)15 = 3	3)12 = 4	9)9 = 1	3)9 = 3
5)18 = 3 (−15, 3)	5)6 = 1 (−5, 1)	9)22 = 2 (−18, 4)	9)40 = 4 (−36, 4)	9)20 = 2 (−18, 2)

3 × 5 = 15	9 × 5 = 45	3 × 1 = 3	3 × 2 = 6	5 × 4 = 20	9 × 3 = 27	3 × 4 = 12

A The big number in a times problem is 45. One small number is 9. What's the third number?
9)45 = 5

B Joe had a large spider collection in his room. Each container held 9 spiders. There were 5 containers on the shelves. How many spiders did Joe have in his collection?
9 × 5 = 45 spiders

C One small number in a times problem is 5. The other small number is 3. What's the third number?
5 × 3 = 15

Lesson 20

Name ____________________

9)18 = 2	3)12 = 4	3)9 = 3	5)25 = 5	5)5 = 1
5)15 = 3	3)15 = 5	1)10 = 10	3)6 = 2	5)10 = 2
9)38 = 4 (−36, 2)	5)20 = 4 (−20, 0)	5)21 = 4 (−20, 1)	9)20 = 2 (−18, 2)	9)29 = 3 (−27, 2)
3)11 = 3 (−9, 2)	5)11 = 2 (−10, 1)	3)12 = 4 (−12, 0)	9)34 = 3 (−27, 7)	9)30 = 3 (−27, 3)

265 × 8 = 2120	934 × 7 = 6538	141 × 6 = 846	794 × 3 = 2382	314 × 5 = 1570	767 × 2 = 1534

27 × 41: 27 + 108 = 1107	38 × 24: 152 + 76 = 912	56 × 37: 392 + 168 = 2072	97 × 11: 97 + 97 = 1067	31 × 31: 31 + 93 = 961

Lesson 22

Name ____________________

5)30 = 6	9)45 = 5	5)35 = 7	3)12 = 4	5)20 = 4
5)10 = 2	5)50 = 10	3)9 = 3	3)15 = 5	9)27 = 3

644 × 8 = 5152	930 × 1 = 930	325 × 8 = 2600	214 × 5 = 1070	931 × 4 = 3724	226 × 4 = 904

9)25 = 2 (−18, 7)	9)32 = 3 (−27, 5)	3)14 = 4 (−12, 2)	9)30 = 3 (−27, 3)	5)9 = 1 (−5, 4)

A Natisha was arranging the desks in her classroom. Each row had 7 desks. If there were 4 rows, how many desks were there?
7 × 4 = 28 desks

B The big number in a times problem is 30. One of the small numbers is 5. What is the other small number?
5)30 = 6

C 2456 people went to see the Mud Buckets play in the park. After the band started playing, 378 more people showed up. How many people went to see the band?
2456 + 378 = 2834 people

Lesson 24   Name _____

$$9\overline{)36}=4 \quad 5\overline{)30}=6 \quad 3\overline{)6}=2 \quad 3\overline{)15}=5 \quad 9\overline{)27}=3$$

$$9\overline{)45}=5 \quad 3\overline{)12}=4 \quad 5\overline{)5}=1 \quad 2\overline{)0}=0 \quad 5\overline{)25}=5$$

$$9\overline{)31}=3,\ -27=4 \quad 5\overline{)13}=2,\ -10=3 \quad 9\overline{)47}=5,\ -45=2 \quad 3\overline{)14}=4,\ -12=2 \quad 5\overline{)18}=3,\ -15=3$$

³¹¹ 4217 −3686 = **531**	²⁹⁹ 3000 −1842 = **1158**	⁵ 3656 −2180 = **1476**	³ 5347 − 319 = **5028**	⁵¹³ 1645 −1188 = **457**

A Shanna sold peanuts at the baseball stadium. She sold each bag for 25 dollars. If Shann sold 9 bags, how much money would she have?

25 × 9 = **225 dollars**

B David loved green beans. One day he ate 43 green beans for lunch. He ate 12 more green beans for dinner than he ate for lunch. How many green beans did David eat for dinner?

43 +12 = **55 green beans**

C The big number in a times problem is 36. One of the small numbers is 9. What is the other small number?

$$9\overline{)36}=4$$

Lesson 26   Name _____

$$7\overline{)0}=0 \quad 5\overline{)20}=4 \quad 3\overline{)6}=2 \quad 9\overline{)18}=2 \quad 2\overline{)0}=0$$

$$3\overline{)12}=4 \quad 5\overline{)30}=6 \quad 9\overline{)45}=5 \quad 9\overline{)36}=4 \quad 5\overline{)25}=5$$

8 ×7 = **56**	3 ×5 = **15**	8 ×5 = **40**	9 ×4 = **36**	5 ×8 = **40**	8 ×6 = **48**	8 ×9 = **72**

⁷ 847 −360 = **487**	¹ 1564 +2241 = **3805**	741 +742 = **1483**	⁵¹³ 3641 −1576 = **2065**	⁹⁹ 1000 − 9 = **991**

326 −215 = **111**	¹¹¹ 2754 +1676 = **4430**	³ 7941 −5412 = **2529**	¹ 3621 −1313 = **2308**	⁴ 544 −450 = **94**

A Richard's farm had 4 barns. Each barn held 9 horses. How many horses did Richard have?

9 × 4 = **36 horses**

B A lemonade stand had regular and cherry lemonade. On Saturday, 245 cups of lemonade were sold. 132 cups were regular. How many cherry lemonades were sold?

245 −132 = **113 cherry lemonades**

Lesson 28   Name _____

$$5\overline{)40}=8 \quad 3\overline{)6}=2 \quad 5\overline{)10}=2 \quad 3\overline{)12}=4 \quad 5\overline{)30}=6$$

$$3\overline{)15}=5 \quad 9\overline{)45}=5 \quad 9\overline{)36}=4 \quad 3\overline{)3}=1 \quad 9\overline{)27}=3$$

2 ×6 = **12**	8 ×5 = **40**	2 ×9 = **18**	3 ×5 = **15**	2 ×7 = **14**	8 ×10 = **80**	2 ×8 = **16**

$$9\overline{)40}=4,\ -36=4 \quad 9\overline{)35}=3,\ -27=8 \quad 3\overline{)11}=3,\ -9=2 \quad 5\overline{)17}=3,\ -15=2 \quad 3\overline{)14}=4,\ -12=2$$

²² 765 × 4 = **3060**	¹¹ 255 × 3 = **765**	¹ 312 × 6 = **1872**	³ 241 × 8 = **1928**	644 × 2 = **1288**	²³ 369 × 4 = **1476**

A A small number in a times problem is 8. The other small number is 7. What is the third number?

8 ×7 = **56**

B The big number in a times problem is 80. One small number is 8. What is the other small number?

$$8\overline{)80}=10$$

Lesson 30   Name _____

$$5\overline{)40}=8 \quad 3\overline{)6}=2 \quad 5\overline{)10}=2 \quad 3\overline{)12}=4 \quad 5\overline{)30}=6$$

$$3\overline{)15}=5 \quad 9\overline{)45}=5 \quad 9\overline{)36}=4 \quad 3\overline{)3}=1 \quad 9\overline{)27}=3$$

8 ×6 = **48**	2 ×5 = **10**	2 ×9 = **18**	8 ×9 = **72**	8 ×8 = **64**	2 ×6 = **12**	2 ×8 = **16**

A 5 kids fit into each car. There were 25 kids. How many cars were there?

$$5\overline{)25}=5 \text{ cars}$$

B 15 kids went to the fair. Each seat on the Ferris wheel held 3 kids. How many seats did the kids fill up?

$$3\overline{)15}=5 \text{ seats}$$

641 −221 = **420**	9456 +1322 = **10,778**	364 +225 = **589**	⁶ 8974 −7829 = **1145**	⁹⁹ 1000 − 432 = **568**

448 −417 = **31**	7416 +4152 = **11,568**	³ 6417 −1241 = **5176**	²¹ 3174 −1212 = **1962**	271 −140 = **131**

Lesson 32 Name ____________

3 5)15 | 6 8)48 | 2 2)4 | 9 8)72 | 5 8)40

5 9)45 | 4 2)8 | 7 5)35 | 6 1)6 | 9 8)72

49 5)245 | 449 8)3596 | 72 3)218 | 1342 5)6714 | 148 3)446

7 × 5 = 35 | 2 × 8 = 16 | 7 × 3 = 21 | 8 × 8 = 64 | 2 × 10 = 20 | 7 × 4 = 28 | 8 × 5 = 40

A Each basket had 9 apples. There were 72 apples. How many baskets were there? 9)72 — 8 baskets

B Every room in Jonah's house had 2 windows. There were 6 rooms. How many windows were there? 2 × 6 — 12 windows

Lesson 34 Name ____________

7 8)56 | 1 5)5 | 6 2)12 | 5 9)45 | 9 8)72

5 3)15 | 2 2)4 | 6 1)6 | 10 5)50 | 5 8)40

558 × 6 = 3348 | 764 × 3 = 2292

986 × 5 = 4930 | 220 × 8 = 1760

328 5)1644 r4 | 2316 2)4632 | 579 5)2896 r1 | 617 8)4941 r5 | 79 3)237

452 − 312 = 140 | 5650 − 2176 = 3474 | 3641 − 2800 = 841 | 7741 − 6854 = 887 | 320 − 310 = 10

Lesson 36 Name ____________

6 8)48 | 2 5)10 | 6 1)6 | 4 3)12 | 5 8)40

9 1)9 | 9 8)72 | 8 5)40 | 5 9)45 | 6 2)12

780 2)1560 | 250 5)1254 r4 | 1205 8)9640 | 32 5)160 | 62 2)124

A Each tree in an apple orchard had 45 apples. There were 12 trees. How many apples were there? 45 × 12 = 90 + 450 — 540 apples

B Every bus leaving the station carried 8 passengers. There were 40 passengers. How many buses were there? 8)40 — 5 buses

C Jason had his baseball cards in plastic bags. Each bag held 9 cards. There were 72 cards. How many bags were there? 9)72 — 8 bags

456 + 544 = 1000 | 6372 − 5895 = 477 | 2644 + 3800 = 6444 | 9897 − 4646 = 5251 | 412 + 258 = 670

Lesson 38 Name ____________

0 5)0 | 2 5)10 | 5 9)45 | 6 2)12 | 2 2)4

7 8)56 | 1 5)5 | 4 3)12 | 5 5)25 | 9 1)9

4 × 6 = 24 | 2 × 9 = 18 | 8 × 5 = 40 | 4 × 8 = 32 | 2 × 6 = 12 | 4 × 5 = 20 | 8 × 9 = 72

334 5)1674 r4 | 1436 2)2872 | 2254 3)6763 r1 | 700 8)5602 r2 | 91 5)455

A 45 airplanes took off each hour. How many airplanes took off in 8 hours? 45 × 8 — 360 airplanes

B 90 airplanes took off from the airport yesterday. If 9 airplanes left each hour, how many hours did it take for all of the airplanes to take off? 9)90 — 10 hours

674 + 580 = 1254 | 7541 + 2261 = 9802 | 3125 + 9754 = 12,879 | 4474 + 7123 = 11,597 | 9745 + 1379 = 11,124

Lesson 40

Name ______________________

$50 \div 10 = 5$ | $28 \div 4 = 7$ | $30 \div 10 = 3$ | $32 \div 4 = 8$ | $36 \div 4 = 9$

$90 \div 10 = 9$ | $25 \div 5 = 5$ | $32 \div 4 = 8$ | $90 \div 10 = 9$ | $45 \div 5 = 9$

$6 \times 1 = 6$ | $10 \times 3 = 30$ | $4 \times 6 = 24$ | $10 \times 8 = 80$ | $6 \times 5 = 30$ | $5 \times 6 = 30$ | $10 \times 10 = 100$

$8\overline{)244}$ = 30 (−24, 04)

$5\overline{)1675}$ = 335 (−15, 17, −15, 25, −25, 0)

$3\overline{)2801}$ = 933 (−27, 10, −9, 11, −9, 2)

$2\overline{)1112}$ = 556 (−10, 11, −10, 12, −12, 0)

$5\overline{)326}$ = 65 (−30, 26, −25, 1)

27×15: 135, 27, 405

32×67: 224, 192, 2144

44×12: 88, 44, 528

67×45: 335, 268, 3015

38×21: 38, 76, 798

Lesson 42

Name ______________________

$32 \div 4 = 8$ | $18 \div 9 = 2$ | $16 \div 2 = 8$ | $27 \div 9 = 3$ | $72 \div 8 = 9$

$18 \div 2 = 9$ | $60 \div 10 = 6$ | $40 \div 5 = 8$ | $56 \div 8 = 7$ | $6 \div 2 = 3$

17×24: 68, 34, 408

38×27: 266, 76, 1026

64×12: 128, 64, 768

36×14: 144, 36, 504

70×40: 2800

A Shannon filed 140 papers in each folder. There were 12 folders. How many papers did Shannon file?

140×12: 280, 1400, 1680 papers

B Janet had 720 flowers in her garden. Each row had 9 flowers. How many rows were there?

$9\overline{)720}$ = 80 rows (−72, 00)

$2765 - 1441 = 1324$ | $6768 + 1111 = 7879$ | $8547 - 3131 = 5416$ | $8754 - 1249 = 7505$ | $112 - 110 = 2$

$6872 + 1333 = 8205$ | $7431 - 3751 = 3680$ | $9526 - 7644 = 2182$ | $7000 + 5 = 7005$ | $200 - 1 = 199$

Lesson 44

Name ______________________

$100 \div 10 = 10$ | $18 \div 9 = 2$ | $50 \div 10 = 5$ | $12 \div 2 = 6$ | $60 \div 10 = 6$

$48 \div 8 = 6$ | $25 \div 5 = 5$ | $15 \div 3 = 5$ | $56 \div 8 = 7$ | $27 \div 9 = 3$

$5\overline{)5150}$ = 1030 (−5, 01, −0, 15, −15, 00)

$8\overline{)2641}$ = 330 (−24, 24, −24, 01)

$3\overline{)6341}$ = 2113 (−6, 03, −3, 04, −3, 11, −9, 2)

$2\overline{)2265}$ = 1132 (−2, 02, −2, 06, −6, 05, −4, 1)

$5\overline{)451}$ = 90 (−45, 01)

$226 \times 4 = 904$ | $38 \times 1 = 38$ | $455 \times 5 = 2275$ | $676 \times 8 = 5408$ | $42 \times 9 = 378$ | $197 \times 4 = 788$

A A shark ate 450 fish. Each hour the shark ate 9 fish. How many hours did it take the shark to eat all of the fish?

$9\overline{)450}$ = 50 hours (−45, 00)

B Randall collected pens. He put 10 pens in each drawer. If Randall had 900 pens, how many drawers had pens in them?

$10\overline{)900}$ = 90 drawers (−90, 00)

Lesson 46

Name ______________________

$18 \div 2 = 9$ | $8 \div 2 = 4$ | $24 \div 4 = 6$ | $35 \div 7 = 5$ | $14 \div 2 = 7$

$32 \div 4 = 8$ | $9 \div 3 = 3$ | $36 \div 9 = 4$ | $40 \div 5 = 8$ | $16 \div 2 = 8$

$7 \times 8 = 56$ | $10 \times 9 = 90$ | $6 \times 5 = 30$ | $7 \times 8 = 56$ | $10 \times 5 = 50$ | $7 \times 7 = 49$ | $6 \times 4 = 24$

A There were 720 people at a concert. 8 people sat in each row. How many rows were there?

$8\overline{)720}$ = 90 rows (−72, 0)

B Mike made 15 dollars per hour. He worked for 40 hours. How much money did he make?

$15 \times 40 = 600$ dollars

C Ron danced for 3 hours every day for one month. There were 30 days in the month. How many hours did Ron dance?

$30 \times 3 = 90$ hours

$6844 + 2876 = 9720$ | $1912 - 5 = 1907$ | $2467 + 3410 = 5877$ | $7641 + 4646 = 12{,}287$ | $2870 - 1805 = 1065$

Lesson 48

Name ____________________

9)45 = 5	5)20 = 4	7)21 = 3	10)90 = 9	7)28 = 4
10)30 = 3	2)6 = 3	10)20 = 2	4)36 = 9	9)72 = 8

4)2745 = 686	2)7860 = 3930	7)542 = 77	5)3221 = 644	4)364 = 91
−24, 34, −32, 25, −24, 1	−6, 18, −18, 06, −6, 00	−49, 52, −49, 3	−30, 22, −20, 21, −20, 1	−36, 04, −4, 0

A Shania had 38 pairs of slacks. She bought 12 more pairs of slacks. How many slacks did Shania end up with?

[ ] ____________

38 + 12 = **50 slacks**

B There were 36 items of clothing in a dresser. Each drawer held 4 items. How many drawers were there?

[ ] ____________

4)36 = **9 drawers**

C A slug ate 2 flowers each hour. The slug ate 464 flowers. How many hours did the slug eat?

[ ] ____________

2)464 = **232 hours**

451 + 212 = 663	7142 − 3800 = 3342	2541 − 1277 = 1264	6766 − 4320 = 2446	9741 + 1000 = 10,741

Lesson 50

Name ____________________

4)32 = 8	2)14 = 7	7)35 = 5	4)24 = 6	2)8 = 4
2)18 = 9	7)14 = 2	10)80 = 8	2)16 = 8	5)40 = 8

1647 + 4645 = 6292	3326 + 4800 = 8126	2674 + 1241 = 3915	9541 + 1212 = 10,753	8547 + 8641 = 17,188

7)265 = 37	5)3846 = 769	9)7541 = 837	3)6786 = 2262	6)722 = 120
−21, 55, −49, 6	−35, 34, −30, 46, −45, 1	−72, 34, −27, 71, −63, 8	−6, 07, −6, 18, −18, 06, −6, 0	−6, 12, −12, 02

5)97,145 = 19,429	3)41,526 = 13,842	6)97,140 = 16,190	4)34,125 = 8531
−5, 47, −45, 21, −20, 14, −10, 45, −45, 0	−3, 11, −9, 25, −24, 12, −12, 06, −6, 0	−6, 37, −36, 11, −6, 54, −54, 00	−32, 21, −20, 12, −12, 05, −4, 1

Lesson 52

Name ____________________

9)45 = 5	5)20 = 4	7)21 = 3	10)90 = 9	7)28 = 4
10)30 = 3	2)6 = 3	10)20 = 2	4)36 = 9	8)72 = 9

9 × 8 = 72	6 × 9 = 54	6 × 6 = 36	4 × 5 = 20	4 × 3 = 12	9 × 7 = 63	6 × 8 = 48

484 × 21	247 × 18	441 × 23	674 × 24	988 × 43
484	1976	1323	2696	2964
968	247	882	1348	3952
10,164	4446	10,143	16,176	42,484

5)8746 = 1749	4)9466 = 2366	7)574 = 82	8)3215 = 401	2)670 = 335
−5, 37, −35, 24, −20, 46, −45, 1	−8, 14, −12, 26, −24, 26, −24, 2	−56, 14, −14, 0	−32, 01, −0, 15, −8, 7	−6, 07, −6, 10, −10, 0

Each box held 9 cartons of paper. There were 279 cartons of paper. How many boxes were there?

[ ] ____________

9)279 = **31 boxes**
−27, 9, −9, 0

Lesson 54

Name ____________________

2)14 = 7	9)18 = 2	10)80 = 8	3)15 = 5	2)6 = 3
8)64 = 8	5)20 = 4	4)24 = 6	2)12 = 6	5)40 = 8

5684 + 2120 = 7804	8716 − 7741 = 975	6412 + 1121 = 7533	4574 − 3147 = 1427	1584 + 2311 = 3895

38 × 76	51 × 26	32 × 18	97 × 41	69 × 36	47 × 12	82 × 46
228	306	256	97	414	94	492
266	102	32	388	207	47	328
2888	1326	576	3977	2484	564	3772

7)67,145 = 9592	4)38,201 = 9550	8)69,124 = 8640	2)52,143 = 26,071
−63, 41, −35, 64, −63, 15, −14, 1	−36, 22, −20, 20, −20, 01	−64, 51, −48, 32, −32, 04	−4, 12, −12, 01, −0, 14, −14, 03, −2, 1

Lesson 56 Name ____________

4)24 = 6	2)12 = 6	5)40 = 8	9)27 = 3	4)36 = 9
3)12 = 4	10)90 = 9	2)18 = 9	8)48 = 6	3)9 = 3

A 6 people could fit into 1 elevator. There were 453 elevators. If all the elevators were full, how many people would they hold?

453 × 6 = 2718 people

B 5765 people went to see the football game. If each row held 5 people, how many rows would be full?

5)5765 = 1153 people
−5, 07, −5, 26, −25, 15, −15

C Every time Janice took a test, she answered 8 problem correctly. If Janice took 31 tests, how many problems would she answer correctly?

31 × 8 = 248 problems

8 × 4 = 32	3 × 6 = 18	9 × 7 = 63	8 × 1 = 8	3 × 9 = 27	9 × 6 = 54	8 × 3 = 24

218 + 641 = 859	3812 − 1595 = 2217	9741 − 2311 = 7430	5841 + 2614 = 8455	9741 − 3813 = 5928

Lesson 58 Name ____________

6)54 = 9	2)2 = 1	8)48 = 6	4)12 = 3	7)21 = 3
8)56 = 7	3)21 = 7	8)72 = 9	9)27 = 3	7)70 = 10

4)4517 = 1129 (−4, 05, −4, 11, −8, 37, −36, 1)

8)6786 = 848 (−64, 38, −32, 66, −64, 2)

5)4316 = 863 (−40, 31, −30, 16, −15, 1)

2)8541 = 4270 (−8, 05, −4, 14, −14, 01)

45)6415 = 142 (−45, 191, −180, 115, −90, 25)

38)1674 = 44 (−152, 154, −152, 2, −0, 2)

52)1147 = 22 (−104, 107, −104, 3)

28)6567 = 234 (−56, 96, −84, 127, −112, 15)

28 × 6 = 168

48 × 48 = 384 + 192 = 2304

19 × 7 = 133

68 × 44 = 272 + 272 = 2992

18 × 7 = 126

46 × 22 = 92 + 92 = 1012

74 × 38 = 592 + 222 = 2812

Lesson 60 Name ____________

9)90 = 10	9)63 = 7	3)6 = 2	3)27 = 9	7)42 = 6
6)54 = 9	2)2 = 1	8)48 = 6	4)12 = 3	7)21 = 3

34)2685 = 78 (−238, 305, −272, 33)

87)7854 = 90 (−783, 24)

13)4812 = 370 (−39, 91, −91, 02)

27)9704 = 359 (−81, 160, −135, 254, −243, 11)

A Jason had 245 more baseball cards than Michael. Michael had 3642 cards. How many baseball cards did Jason have?

3642 + 245 = 3887 cards

B A supermarket paid a local television station 4758 dollars for television commercials. 6 commercials were made. How much did each commercial cost?

6)4758 = 793 dollars
−42, 55, −54, 18, −18

C There were 2457 cookies made for the school bake sale. 9 people helped make them. How many cookies did each person make?

9)2457 = 273 cookies
−18, 65, −63, 27, −27

Lesson 62 Name ____________

2)18 = 9	8)72 = 9	3)24 = 8	3)12 = 4	6)18 = 3
8)64 = 8	7)63 = 9	7)70 = 10	4)12 = 3	3)27 = 9

9714 − 1241 = 8473	4714 − 1675 = 3039	9741 − 1214 = 8527	1347 − 655 = 692	8648 − 4613 = 4035
1946 + 2417 = 4363	3746 + 9321 = 13,067	2858 + 1369 = 4227	2855 + 3639 = 6494	4656 + 2312 = 6968

3)8765 = 2921 (−6, 27, −27, 06, −6, 05, −3, 2)

8)2683 = 335 (−24, 28, −24, 43, −40, 3)

3)2212 = 737 (−21, 11, −9, 22, −21, 1)

7)3612 = 516 (−35, 11, −7, 42, −42, 0)

Lesson 62 (continued)   Name ____________

- 13 ⟌ 4512 = 347 — −39, 61, −52, 92, −91, 1
- 17 ⟌ 2107 = 123 — −17, 40, −34, 67, −51, 16
- 4 ⟌ 6822 = 1705 — −4, 28, −28, 02, −0, 22, −20, 2
- 8 ⟌ 2412 = 301 — −24, 01, −0, 12, −8, 4
- 9 ⟌ 455 = 50 — −45, 05, −0, 5
- 22 ⟌ 6912 = 314 — −66, 31, −22, 92, −88, 4
- 8 ⟌ 108 = 13 — −8, 28, −24, 4
- 91 ⟌ 5760 = 63 — −546, 300, −273, 27

Lesson 64   Name ____________

- 6 ⟌ 30 = 5
- 8 ⟌ 72 = 9
- 2 ⟌ 16 = 8
- 3 ⟌ 27 = 9
- 7 ⟌ 21 = 3
- 9 ⟌ 9 = 1
- 3 ⟌ 12 = 4
- 8 ⟌ 48 = 6
- 7 ⟌ 70 = 10
- 2 ⟌ 20 = 10
- 8 ⟌ 2765 = 345 — −24, 36, −32, 45, −40, 5
- 6 ⟌ 55,874 = 9312 — −54, 18, −18, 07, −6, 14, −12, 2
- 3 ⟌ 2128 = 709 — −21, 02, −0, 28, −27, 1
- 72 ⟌ 4484 = 62 — −432, 164, −144, 20
- 62 ⟌ 482 = 7 — −434, 48
- 38 ⟌ 4712 = 124 — −38, 91, −76, 152, −152, 0
- 5 ⟌ 24,150 = 4830 — −20, 41, −40, 15, −15, 0
- 67 ⟌ 3820 = 57 — −335, 470, −469, 1

Lesson 64 (continued)   Name ____________

A A garden of tomato plants grew 476 tomatoes. Each tomato plant grew 7 tomatoes. How many tomato plants were there?

☐ __________

7 ⟌ 476 = 68 plants — −42, 56, −56, 0

B Ken flew on 2895 airplanes. Every week he flew on 5 airplanes. How many weeks did Ken fly on airplanes?

☐ __________

5 ⟌ 2895 = 579 weeks — −25, 39, −35, 45, −45, 0